The Prevention of Cardiovascular Disease Through the Mediterranean Diet

The Prevention of Cardiovascular Disease Through the Mediterranean Diet

Almudena Sánchez-Villegas
Professor of Preventive Medicine and Public Health,
University of Las Palmas de Gran Canaria

Ana Sánchez-Tainta
RD, Registered Dietitian, Department of Preventive Medicine and
Public Health, University of Navarra, Pamplona, Spain

ACADEMIC PRESS
An imprint of Elsevier

Academic Press is an imprint of Elsevier
125 London Wall, London EC2Y 5AS, United Kingdom
525 B Street, Suite 1800, San Diego, CA 92101-4495, United States
50 Hampshire Street, 5th Floor, Cambridge, MA 02139, United States
The Boulevard, Langford Lane, Kidlington, Oxford OX5 1GB, United Kingdom

Notices
Knowledge and best practice in this field are constantly changing. As new research and experience broaden
our understanding, changes in research methods, professional practices, or medical treatment may become
necessary.

Practitioners and researchers must always rely on their own experience and knowledge in evaluating and
using any information, methods, compounds, or experiments described herein. In using such information or
methods they should be mindful of their own safety and the safety of others, including parties for whom they
have a professional responsibility.

To the fullest extent of the law, neither the Publisher nor the authors, contributors, or editors, assume any
liability for any injury and/or damage to persons or property as a matter of products liability, negligence or
otherwise, or from any use or operation of any methods, products, instructions, or ideas contained in the
material herein.

British Library Cataloguing-in-Publication Data
A catalogue record for this book is available from the British Library

Library of Congress Cataloging-in-Publication Data
A catalog record for this book is available from the Library of Congress

ISBN: 978-0-12-811259-5

For Information on all Academic Press publications
visit our website at https://www.elsevier.com/books-and-journals

Working together
to grow libraries in
developing countries

www.elsevier.com • www.bookaid.org

Publisher: Mica Haley
Acquisition Editor: Stacy Masucci
Editorial Project Manager: Sam Young
Production Project Manager: Chris Wortley & Priya Kumaraguruparan
Cover Designer: Matthew Limbert

Typeset by MPS Limited, Chennai, India

Contents

List of Contributors   xiii
Foreword   xv

1. **A Healthy-Eating Model Called Mediterranean Diet**
   Almudena Sánchez-Villegas and Itziar Zazpe

   1.1 Definitions and History of the Mediterranean Diet   1
   1.2 Mediterranean Diet: Characteristics   2
   1.3 Mediterranean Diet Pyramids   4
   1.4 Nutritional Composition and Nutritional Adequacy
       of the Mediterranean Diet   8
   1.5 How to Assess the Adherence to the Mediterranean Diet   11
   1.6 The 2015 Dietary Guidelines for Americans and the
       Healthy Dietary Patterns Such as the Mediterranean Diet   13
   1.7 Epidemiological Evidences Regarding the Beneficial Role
       of the Mediterranean Diet in Cardiovascular Disease
       and Cardiovascular Risk Factors   14
   1.8 Conclusions   21
   References   21

2. **Epidemiological and Nutritional Methods**
   Estefanía Toledo

   2.1 How Do We Classify the Exposure in Nutritional
       Epidemiology?   25
   2.2 How Do We Measure the Occurrence of Disease?   26
       2.2.1 Prevalence Proportion   26
       2.2.2 Prevalence Odds   26
       2.2.3 Cumulative Incidence   27
       2.2.4 Incidence Rate or Incidence Density   27
       2.2.5 Hazard Function   28
   2.3 How Do We Assess If a Certain Exposure Is Associated
       to a Certain Outcome?   28
       2.3.1 Relative Risk   28
       2.3.2 Odds Ratio   29
       2.3.3 Hazard Ratio   30
   2.4 What Type of Studies Do We Use in Epidemiology?   30
       2.4.1 Ecological Studies   31

2.4.2	Cross-Sectional Studies	32
2.4.3	Case-Control Studies	32
2.4.4	Cohort Studies	32
2.4.5	Randomized Controlled Trials	33
2.4.6	Meta-Analysis	34
References		34

3.   Not All Fats Are Unhealthy

Ligia J. Dominguez and Mario Barbagallo

3.1	Introduction	35
3.2	What Is Fat?	37
3.3	Fat Absorption	42
3.4	Good Fats	44
	3.4.1 Monounsaturated Fats	44
	3.4.2 Polyunsaturated Fats	47
3.5	Bad Fats	48
	3.5.1 Trans Fats	48
3.6	In-Between Fats	51
	3.6.1 Saturated Fats	51
3.7	All Types of Fats Are Part of Foods and Dietary Patterns	52
References		55

4.   Virgin Olive Oil: A Mediterranean Diet Essential

Almudena Sánchez-Villegas and Ana Sánchez-Tainta

4.1	Introduction	59
4.2	Composition of Olive Oil	59
4.3	Natural Juice of the Olive, Why Extra-Virgin?	60
4.4	Effects on Health	61
	4.4.1 Effects on Triglycerides, LDL Cholesterol, and HDL Cholesterol	61
	4.4.2 Antioxidant and Antiinflammatory Effect	63
	4.4.3 Effect on Endothelial Function	64
	4.4.4 Antithrombotic Effect	65
4.5	Epidemiological Evidences Regarding the Role of Olive Oil in Cardiovascular Disease	66
4.6	Epidemiological Evidences Regarding the Association Between Olive Oil Consumption and Cardiovascular Risk Factors	74
	4.6.1 Diabetes	74
	4.6.2 Hypertension	79
	4.6.3 Obesity	81
	4.6.4 Metabolic Syndrome	81
4.7	How Much Oil Is Recommended a Day?	82
4.8	Consumer Tips: How to Take Advantage of Olive Oil	82
4.9	How Can I Incorporate Olive Oil in My Diet?	83
References		83

5. A Small Handful of Mixed Nuts

Maira Bes-Rastrollo and Ana Sánchez-Tainta

5.1 Introduction 89
5.2 Scientific Evidence 90
    5.2.1 The Secret of Nuts: Their Nutrient Content 90
    5.2.2 Epidemiological Evidence on Healthy Effects of Nut Consumption: Observational Studies 92
    5.2.3 Epidemiological Evidence on Healthy Effects of Nut Consumption: Clinical Trials 94
    5.2.4 Eating Nuts Does Not Make You Fat! 95
5.3 Recommendations 96
    5.3.1 Consumption Advice 96
    5.3.2 How to Introduce Nuts in Our Diets 96
References 97

6. Fruits and Vegetables

Angeliki Papadaki and Ana Sánchez-Tainta

6.1 Introduction 101
6.2 Consumption of Fruits and Vegetables and Risk of Cardiovascular Disease 102
6.3 Consumption of Fruits and Vegetables and Cardiovascular Risk Factors 103
    6.3.1 Obesity 103
    6.3.2 Hypertension 104
    6.3.3 Type 2 Diabetes 105
6.4 How Much Should I Eat? 105
6.5 What Counts as a Serving? 106
6.6 How Can I Increase My Fruit and Vegetable Consumption? 107
6.7 Summary and Recommendations 107
References 108

7. Cereals and Legumes

Karen J. Murphy, Iva Marques-Lopes and Ana Sánchez-Tainta

7.1 Cereals and Cardiovascular Disease 111
    7.1.1 Overview and Introduction of Cereals in the Mediterranean Diet 111
    7.1.2 Whole-Grain Cereals and Cardiovascular Disease: Epidemiological Evidence 112
    7.1.3 Biological Mechanisms of Benefit of Whole Grains 115
    7.1.4 Conclusions 116
    7.1.5 Cereal Consumption Recommendations 116

7.2   Legumes and Cardiovascular Disease   117
      7.2.1   Overview and Introduction of Legumes in the
              Mediterranean Diet   117
      7.2.2   Legumes and Cardiovascular Disease: Epidemiological
              Evidence   118
      7.2.3   Nutritional Composition, Dietary Fibers, and
              Phytochemicals of Legumes: Their Role on
              Cardiovascular Risk   119
7.3   Conclusions   126
7.4   Legume Consumption Recommendations   127
References   128

8.   More Fish, Less Meat

Mary K. Downer and Ana Sánchez-Tainta

8.1   Early Research on Fish Intake and Cardiovascular
      Disease   133
8.2   Nutritional Composition of Fish in Relation to
      Cardiovascular Benefits   133
8.3   Nutritional Composition of Meat in Relation to
      Cardiovascular Benefits   134
8.4   Intermediate Physiological Effects of Fish Intake   135
      8.4.1   Lowering Plasma Triglycerides   135
      8.4.2   Reducing Heart Rate and Blood Pressure   136
      8.4.3   Improving Myocardial Filling and Efficiency   136
      8.4.4   Decreasing Inflammation   137
      8.4.5   Antiarrhythmia Effects   137
      8.4.6   Visceral Adiposity   137
8.5   Intermediate Physiological Effects of Meat Intake   137
      8.5.1   Increased LDL: HDL Serum Cholesterol
              Ratio   138
      8.5.2   Increased Triglycerides   138
      8.5.3   Insulin Resistance, Hyperglycemia,
              Hyperinsulinemia   138
      8.5.4   Body Weight   138
      8.5.5   Worse Arterial Compliance and Vascular
              Stiffness   139
      8.5.6   Hypertension   139
8.6   Epidemiological Evidence—Fish Intake and
      Cardiovascular Disease   139
8.7   Epidemiological Evidence—Meat Intake and
      Cardiovascular Disease   141
8.8   Fish, Mercury, and Cardiovascular Disease   142
8.9   Conclusions   143
8.10 Recommendations   143
References   144

9. Red Wine Moderate Consumption and at Mealtimes

Alfredo Gea and Ana Sánchez-Tainta

9.1	Introduction	151
9.2	Scientific Evidence	152
	9.2.1 Mortality	152
	9.2.2 Cardiovascular Diseases	153
	9.2.3 Diabetes	154
	9.2.4 Depression	154
	9.2.5 Weight	154
	9.2.6 Cancer	154
	9.2.7 Other Health Effects	155
9.3	Recommendations	155
	References	156

10. The Mediterranean Lifestyle: Not Only Diet But Also Socializing

Ignacio Ara

10.1	Introduction	159
10.2	Exploring Cardiovascular Risk Factors in the Mediterranean Countries: Potential Consequences	160
10.3	Climate Conditions and Opportunities for a Healthy Lifestyle	161
10.4	Why Can All These Climate Conditions Be of Interest When Studying Cardiovascular Risk?	161
10.5	Mental Well-Being: The Importance of Socialization	163
10.6	Future Perspectives: Urban Environments and Their Relationship to Food Consumption and an Active Lifestyle	165
	References	166

11. A Healthy Diet for Your Heart and Your Brain

Almudena Sánchez-Villegas and Elena H. Martínez-Lapiscina

11.1	Mediterranean Diet and Dementia	169
	11.1.1 Dementia: A Public Health Priority	169
	11.1.2 Dementia Is Incurable but Preventable: Relevance of Pathogenic Mechanisms	170
	11.1.3 Dementia: Hope Through Mediterranean Diet	171
	11.1.4 Epidemiological Evidences Regarding the Role of Mediterranean Diet in Cognitive Decline and Dementia	172
	11.1.5 Ongoing Efforts: From Dietary Toward Holistic Interventions	180
	11.1.6 Recommendations	181

11.2   Mediterranean Diet and Depression   182
 11.2.1 Diet, a New Target to Prevent Depression   182
 11.2.2 Etiological Hypothesis of Depression   182
 11.2.3 Depression, Cardiovascular Risk Factors, and
         Cardiovascular Disease   184
 11.2.4 Biological Plausibility for the Link Between Cardio-
         Protective Mediterranean Diet and Reduced Risk
         of Depression   185
 11.2.5 Epidemiological Evidences Regarding the Role of
         Mediterranean Diet in Depression   185
 11.2.6 Recommendations   191
 References   192

12.   The Mediterranean Cook: Recipes for All Seasons
María Soledad Hershey and Ana Sánchez-Tainta

12.1   Salads and Vegetables   200
 12.1.1 Orange and Raisin Endive Salad   200
 Directions   200
 12.1.2 Feta Cheese and Nut Spinach Salad   200
 Directions   200
 12.1.3 Cream of Butternut Squash   201
 Directions   201
 12.1.4 Tomato Soup   201
 Directions   201
 12.1.5 "Escalibada" Toast   201
 Directions   202
12.2   Pasta and Rice   202
 12.2.1 Pasta With Red Peppers and Tuna   202
 Directions   202
 12.2.2 Pea and Mushroom Spaghetti   202
 Directions   203
 12.2.3 Whole-Grain Noodles With Mussels   203
 Directions   203
 12.2.4 Vegetable Soup With Brown Rice   203
 Directions   204
 12.2.5 Seafood Paella   204
 Directions   204
12.3   Beans   204
 12.3.1 Chickpea and Egg Salad   204
 Directions   205
 12.3.2 Chickpeas With Codfish   205
 Directions   205
 12.3.3 Mediterranean Lentils   206
 Directions   206
 12.3.4 Greek "Fasolada"   206
 Directions   206
 12.3.5 Sautéed Peas   207
 Directions   207

12.4 Egg   207
  12.4.1 Spanish Omelet With "Pisto"   207
  Directions   208
  12.4.2 Broccoli and Cheese Omelet   208
  Directions   208
  12.4.3 Scrambled Eggs With Asparagus   208
  Directions   209
  12.4.4 Poached Eggs With Clams and Marinara Sauce   209
  Directions   209
  12.4.5 Fried Eggs With Mushrooms   210
  Directions   210
12.5 Fish   210
  12.5.1 Sole With Clams   210
  Directions   211
  12.5.2 Squid Skewers   211
  Directions   211
  12.5.3 Codfish Served With Onion and Tomato Sauce   211
  Directions   212
  12.5.4 Grilled Salmon With Rice   212
  Directions   212
  12.5.5 Swordfish With Tomato and Black Olives   213
  Directions   213
12.6 Meats   213
  12.6.1 Baked Chicken With Vegetables   213
  Directions   213
  12.6.2 Chicken Thighs With Olives   214
  Directions   214
  12.6.3 Turkey Cooked With Tomatoes and Peppers   214
  Directions   215
  12.6.4 Turkey With Sautéed Zucchini and Onion Sandwich   215
  Directions   215
  12.6.5 Pork Tenderloin With Apple Sauce   215
  Directions   215

Index   217

List of Contributors

Ignacio Ara University of Castilla-La Mancha (UCLM), Ciudad Real, Spain

Mario Barbagallo University of Palermo, Palermo, Italy

Maira Bes-Rastrollo University of Navarra, Pamplona, Spain; Instituto de Salud Carlos III (ISCIII), Madrid, Spain; Instituto de Investigación Sanitaria de Navarra (IdiSNA), Pamplona, Spain

Ligia J. Dominguez University of Palermo, Palermo, Italy

Alfredo Gea University of Navarra, Pamplona, Spain

María Soledad Hershey University of Navarra, Pamplona, Spain

Iva Marques-Lopes University of Zaragoza, Zaragoza, Spain

Elena H. Martínez-Lapiscina Hospital Clinic of Barcelona and Institut d'Investigacions Biomèdiques August Pi Sunyer (IDIBAPS), Barcelona, Spain

Mary K. Downer Harvard TH Chan School of Public Health, Boston, MA, United States; Brigham & Women's Hospital, Boston, MA, United States

Karen J. Murphy University of South Australia, Adelaide, SA, Australia

Angeliki Papadaki University of Bristol, Bristol, United Kingdom

Ana Sánchez-Tainta University of Navarra, Pamplona, Spain; Brigham & Women's Hospital, Boston, MA, United States; Instituto de Salud Carlos III (ISCIII), Madrid, Spain; Instituto de Investigación Sanitaria de Navarra (IdiSNA), Pamplona, Spain

Almudena Sánchez-Villegas University of Las Palmas Gran Canaria, Las Palmas, Spain; Instituto de Salud Carlos III (ISCIII), Madrid, Spain

Estefanía Toledo University of Navarra, Pamplona, Spain

Itziar Zazpe University of Navarra, Pamplona, Spain; Instituto de Salud Carlos III (ISCIII), Madrid, Spain

Foreword

With the last 10 years of publications on the Mediterranean Diet (MD), there is a need to review where the research on the prevention of cardiovascular disease through the MD stands today. For this endeavor, there is no better choice to write a book on this topic than those who have been involved with the SUN and the PREDIMED studies.

This is the case with this book entitled *The Prevention of Cardiovascular Disease Through the Mediterranean Diet* by Almudena Sánchez-Villegas and Ana Sánchez-Tainta, which includes chapters quoting the SUN cohort and the PREDIMED study, to date the most internationally known MD intervention trial for primary prevention of cardiovascular disease. These chapters not only cover a vast array of topics dealing with the role of characteristic foods in the MD, but also provide information on the epidemiological and nutritional methods as well as the wider concepts of the MD in which social aspects are presented. The authors deal with all the above issues demonstrating clearly the impact on the prevention of cardiovascular disease through the MD and offer the readers a broader perspective.

In purely descriptive terms the traditional MD is the dietary pattern prevailing among the people of the olive tree-growing areas of the Mediterranean basin before the mid-1960s, which is before globalization made its influence on lifestyle, including diet.

The recorded history of the traditional MD as a health promoting diet started with the work in the 1960s of Ancel Keys and the legendary Seven Countries Study in which the MD was considered as a low saturated lipid diet that was conveying protection against coronary heart disease by lowering plasma cholesterol levels. Indeed, after the Second World War, it had been observed that the inhabitants of Crete (Greece's largest island) and South Italy had a surprisingly low cardiovascular mortality rate lower than that observed in the United States. This led Dr. Ancel Keys and his colleagues to undertake the Seven Countries Study. In this ecological study the authors collected information on rates of coronary disease, and lifestyles including diet of different populations from seven countries from the northern and southern Europe, the United States, and Japan. In this study, not only significant differences in dietary habits between Japan, the United States, and northern and southern Europe (represented by Finland, Greece, the Netherlands and Italy, and Yugoslavia) were found, but also a strong correlation between saturated fatty acids intake and incidence and mortality of coronary heart disease. After 25 years of follow-up, compared with the value observed in the United States and northern Europe as well as in other areas of southern Europe, coronary heart disease mortality in Crete was strikingly low. The healthy characteristics of the diet of the Greeks, especially of the Cretans, moved Keys to describe this diet as "Mediterranean Diet." The characteristics of the Cretan diet had previously been described and published as a monograph in 1953 entitled *Crete: A Case Study of an Underdeveloped Area*. In this monograph, it was very relevant that dietary habits in this Greek island had remained almost unchanged in the last 40 centuries and were based on a large consumption of cereals, legumes, fruits, vegetables, and olives, along with a limited amount of goat meat, milk, meat products, fish, and wine. In addition, it was reported that a meal was not considered complete without the presence of bread; and that olives and olive oil contributed greatly to energy intake. Regarding oil consumption, it was reported that so much olive oil was used, that food appeared to be immersed "literally" in oil.

In the 1980s the interest in the MD as an integral entity, rather than as a blood cholesterol-lowering diet, was renewed and, in 1995 a simple score assessing conformity to the salient characteristics of the MD was introduced. This score or its variants have allowed the evaluation in cohort and case—control studies of the relation between conformity to this diet and several outcomes, including total mortality, incidence of, or mortality from, cardiovascular diseases as well as incidence of, or mortality from, cancer overall or from specific cancer sites. Recently the PREDIMED study, a MD intervention trial for primary prevention of cardiovascular disease, has become a landmark in the history of MD research as it obtains high-level scientific evidence on the role of MD in the prevention of cardiovascular diseases.

Today it is known that the MD, as a dietary pattern, is characterized not only by a low saturated fats intake capable of reducing blood cholesterol levels, but also by an abundant consumption of virgin olive oil and other components and molecular or biochemical mediators with beneficial effects on health. In this regard, epidemiological studies carried out in the last decades have achieved in getting the MD included as a possible healthy dietary pattern for the reduction of cardiovascular risk in the latest American dietary guidelines.

The MD has been documented to be a very healthy diet. The health attributes of this diet could be partly attributed to traditional foods, which this diet incorporates. The traditional MD, however, is more than just a diet; it is a whole healthy lifestyle pattern and a valuable cultural heritage of people and cultural exchange for millennia, representing much more than just a nutritional, tasty, and healthy diet. It is a balanced lifestyle that includes recipes, cooking methods, celebrations, customs, typical products, and various human activities. For these reasons the MD was acknowledged by UNESCO in 2010 as an Intangible Cultural Heritage.

Despite being widely documented and acknowledged as a healthy diet the MD is paradoxically becoming less the diet of choice in most Mediterranean countries. Indeed, in recent decades due to the progressive spread of the westernized economy, which includes determinants such as tourism, urbanization, or technological development, as well as the globalization of food production and consumption where local, fresh, and seasonal products are being replaced by products with an important ecological and economic footprint.

It is, therefore, a priority for governments and individuals, especially in the Mediterranean area, to preserve and promote this cultural and healthy heritage for generations to come.

Prof. Antonia Trichopoulou
President, Hellenic Health Foundation

Chapter 1

A Healthy-Eating Model Called Mediterranean Diet

Almudena Sánchez-Villegas and Itziar Zazpe

1.1 DEFINITIONS AND HISTORY OF THE MEDITERRANEAN DIET

The Mediterranean Diet is a widely used concept that has been used over the last few decades by doctors, researchers, nutritionists, dietitians, and experts from various fields, even though there is no clear consensus on its definition.

The healthy-eating model called the Mediterranean Diet dates back to the early 1960s, when Keys et al. traveled through southern European countries and started the Seven Countries Study. This study, which was the first piece of research into the Mediterranean Diet, began in 1958 and demonstrated that the mortality rate from coronary heart disease in southern Europe was two to three times lower than in northern Europe or the United States [1,2]. It also described low rates of coronary heart disease in the Mediterranean regions when contrasted with other study populations [2,3], and was the main source of information regarding traditional Mediterranean diets in the 1960s and 1970s [4].

In this context, the Mediterranean Diet was defined as the eating habits observed in Greece and Southern Italy in the late 1950s and early 1960s, when the effects of World War II had passed but the fast-food culture had not yet invaded the area [5].

Seen from another point of view the Mediterranean Diet could be also defined by its clearly beneficial health effects. The first results of the Seven Countries Study showed that there was an association between this traditional dietary pattern and a low incidence, and prevalence of mortality rates from coronary heart diseases, other cardiovascular diseases, and low all-cause mortality, including cancer [6].

In general, the term Mediterranean Diet refers to dietary patterns found in the olive tree-growing areas of the Mediterranean basin before the mid-1960s [7].

The Prevention of Cardiovascular Disease Through the Mediterranean Diet.
DOI: http://dx.doi.org/10.1016/B978-0-12-811259-5.00001-9

Although this concept is a clear combination of history, culture, and environment and implies a common dietary pattern in Mediterranean countries, there are however important differences in typical products, food cultures, traditions, and geographical and ecological environments between regions [8].

From these various approaches, it can be said that there is no single Mediterranean Diet and that there are major differences in the traditional eating habits of citizens from countries bordering the Mediterranean Sea: Italy, Greece, France, Spain, North Africa, and the Eastern basin [9].

Even the dietary practices within the same country can vary considerably. Notwithstanding these differences, however, olive oil has always had a central position in all Mediterranean countries or regions [10].

A recent review considers the Mediterranean Diet as an intangible and sustainable food culture transmitted from generation to generation over centuries, of landscapes, places, knowledge, know-how, technologies, products, food preparation and intake, myths and beliefs, accents, creativity, and hospitalities [11,12]. Indeed, since November 16, 2010, the Mediterranean Diet has been inscribed into UNESCO's Representative List of Intangible Cultural Heritage of Humanity [13]. The objective of this initiative was to safeguard the immense legacy representing the cultural value of the Mediterranean Diet, as well as to share and disseminate its values and benefits internationally.

In conclusion, the Mediterranean Diet is essentially part of a lifestyle and requires the simultaneous consideration of other nondietary behavioral factors when assessing its beneficial health effects [14]. Unfortunately, over the last decades, current diets in Mediterranean countries are departing from the traditional Mediterranean Diet toward an unhealthier eating model. This is due to the widespread dissemination of Western-type culture, along with the globalization of food production and consumption, which is related to the homogenization of food behavior in the modern era [8,15].

Therefore, it may now be time to update the definition of the Mediterranean Diet [16,17].

1.2 MEDITERRANEAN DIET: CHARACTERISTICS

In the mid-20th century Mediterranean peoples frequently consumed high amounts of plant foods (olive oil, fruits, vegetables, legumes, fruits, bread, cereal products, nuts, and seeds), food from animals in limited amounts (fish, meat, and milk and milk products), wine, and had a low intake of saturated fats [6].

Taking into account all these characteristics and using a somewhat reductionist approach, this dietary pattern can be considered to be mainly, but not dogmatically, an exclusively plant-based dietary pattern. However, it has been pointed out that the Mediterranean Diet is not a vegetarian diet, in spite of the fact that two of its most common and typical foods, olive oil and wine, are plant products [14].

Therefore, the traditional Mediterranean Diet favored the consumption of locally grown, seasonally fresh and minimally processed foods and was consumed by physically active people [10].

The main and typical foods of the traditional diet of Mediterranean countries are:

- **Vegetables and fruits.** These are important sources of minerals, vitamins, antioxidants, and fiber. Furthermore, the benefits of their consumption are amplified if they are cooked or dressed with olive oil. Other healthy vegetable options are raw foods such as salads.
- **Grains.** The nutritional composition of grains may vary depending on the variety and environmental growing conditions. In general, cereal grains are high in carbohydrates, low in fat, good sources of protein and provide varying amounts of fiber, vitamins, and minerals. Cereal products should contain whole grains, including wheat, oats, rice, rye, barley, and corn. Grains should also be consumed in minimally processed forms.
- **Olives and olive oil.** By definition, olive oil is a central component in the cuisine of the countries surrounding the Mediterranean Sea. This type of vegetable oil has a peculiar fatty acid composition (with a large proportion of monounsaturated fat—mainly oleic acid—and a relatively low proportion of saturated fat) and also contains other minor compounds (tocopherols and carotenoids among others) with antioxidant properties. Olive oil is the principal source of dietary fat used to dress salads and vegetables, and in cooking or baking. The variety "extra-virgin" olive oil is highest in health promoting fats, phytonutrients, and other important micronutrients.
- **Nuts, legumes, and seeds.** All of these foods are packed with vitamins and minerals. Nuts and seeds also provide healthy mono- and polyunsaturated plant oils as well as protein. Legumes, which include beans, are filling and also contain lean protein.
- **Fish and shellfish** are preferred over meat in the traditional Mediterranean diet, although the amount of fish consumed varies widely between and within Mediterranean countries. This group of foods is an important source of healthy protein and essential heart-healthy omega-3 fatty acids.
- **Cheese and yogurt.** Dairy products, from a variety of animals, principally in the form of yogurt and cheese, are consumed in low to moderate amounts.
- **Eggs** are an excellent source of high-quality protein and also contain a number of healthy nutrients, including B vitamins and protein.
- **Meat.** In general, the consumption of red meat and processed meats is lower in the Mediterranean population than the consumption of white meat, in spite of the fact that red meat is a good source of animal protein.

- **Wine.** Moderate wine drinking in the context of the meals has been a long-standing tradition in the Mediterranean basin, with the exception of Islamic populations of this area.
- **Herbs and spices** add flavors and aromas to foods, reducing at the same time the need to add salt or fat when cooking. Herbs and spices are very common in Mediterranean cuisine; they contain several health-promoting antioxidants and contribute to the differences between the broad varieties of culinary cultures.

1.3   MEDITERRANEAN DIET PYRAMIDS

The first pyramid representing the Mediterranean Diet was published in 1995 (Fig. 1.1), in the context of the International Conference on the Diets of the Mediterranean organized by Oldways Preservation & Exchange Trust and

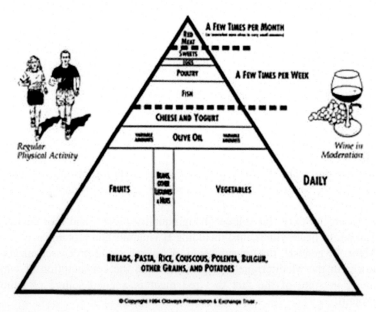

FIGURE 1.1   The Mediterranean Diet Pyramid: a cultural model for healthy eating.

the World Health Organization (WHO)/Food and Agriculture Organization (FAO) in collaboration with the Harvard T.H. Chan School of Public Health [6]. One objective of the group of international experts on diet was to develop a pyramid that summarized the Mediterranean diets consumed during the previous half century.

This pyramid did not specify the relative frequency intake of foods or the proportion of energy obtained from them and in line with other pyramids it was divided into three levels of recommended consumption: daily (cereals and their derived foods, potatoes; fruits, vegetables, legumes and nuts, olive oil in varying amounts, and cheese and yogurt), a few times per week (fish, poultry, eggs, and sweet foods) and a few times per month (red meat). Physical activity and wine consumption in moderation and with meals were two main lifestyle factors which were typical of the Mediterranean areas in the early 1960s, and which were included in the pyramid. Moreover, certain lifestyle factors were considered to be noteworthy: delicious meals, carefully cooked food, eating in friendly company, lengthy meals, and postlunch siestas that provide relaxation.

Although in 2000 the Oldways Preservation & Exchange Trust published yet another pyramid of the Mediterranean Diet, it was not until 2009 when a new, updated and more colorful pyramid was created (Fig. 1.2) [18].

The team of scientists in charge created an entirely new pyramid graphic to better reflect the delicious and appetizing nature of typical Mediterranean foods. The changes focused on gathering plant foods (fruits, vegetables, grains—mostly whole grains—nuts, legumes, beans, seeds, olives and olive oil, and herbs and spices) into a single group to visually emphasize the importance of healthy plant foods in this health-promoting eating pattern. Another important change was the inclusion of fish and shellfish into the pyramid, recognizing the benefits of eating fish and shellfish at least twice a week.

The Mediterranean Diet Foundation in collaboration with the Forum on Mediterranean Food Cultures, Centre International de Hautes Etudes Agronomiques Méditerranéennes, Centro Interuniversitario Internazionale di Studi sulle Culture Alimentari Mediterranee, Sapienza University of Rome, Hebrew University, Jerusalem, and International Commission on the Anthropology of Food and Nutrition worked to develop a revised pyramid of this dietary pattern between 2009 and 2010, based on the latest scientific evidence on nutrition and health (Fig. 1.3). These new dietary guidelines adapted to a new way of life, have taken into account not only quantities, but also the frequency recommendations for the choice of food groups. "The new Mediterranean Diet Pyramid: a lifestyle for today" established daily, weekly, and occasional dietary recommendations [19].

The main recommendations included in this pyramid with regard to daily consumption were:

- The dietary intake should be divided into three main meals and contain three basic foods: cereals, vegetables, and fruit.
- A daily intake of 1.5−2 L of water.

FIGURE 1.2   Mediterranean Diet Pyramid: a contemporary approach to delicious, healthy eating.

- Low fat-dairy products (yogurt, cheese, and other fermented dairy products).
- Olive oil as the principal source of fat in the diet.
- Spices, herbs, onions, and garlic to improve the flavor and palatability of the diet.
- Olives, nuts, and seeds, which can be eaten as healthy snack choices.
- Moderate consumption of wine and other fermented beverages during meals (one glass per day for women, and two glasses for men).

In the middle of the pyramid the following food sources, which are a variety of plant and animal proteins, and which should be consumed weekly, were represented:

- **Legumes** (more than two servings) **and potatoes** (three or fewer servings) per week, preferably fresh potatoes as an accompaniment to fish or meat.

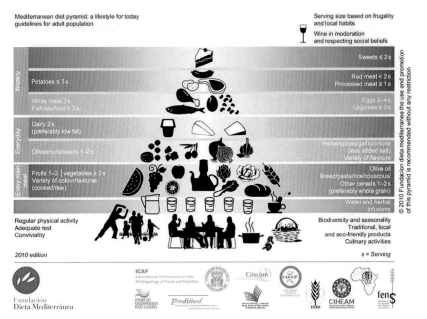

FIGURE 1.3   Mediterranean Diet Pyramid: a lifestyle for today.

- **Fish and white meat**, two or more servings per week, **eggs** (two to four servings).
- **Red meat** (less than two servings, giving preference to lean cuts) and **processed meats** (less than one serving). Both foods should be consumed infrequently and in small quantities.

At the apex of the pyramid, there was one single category:

- **Sweet foods** (sugar, candies, pastries, sweetened fruit juices, and soft drinks). All these foods should be consumed only on special occasions.

    Finally, the newest message in this pyramid was the inclusion of the following lifestyle and cultural elements:

- **Moderation** in portion sizes and always adapted to the energy requirements of each person.
- **Cooking** one's own meals, avoiding processed meals, can be relaxing and fun.
- **The preference for seasonal, fresh, and minimally processed foods** contributes to maximizing the nutritional content of protective nutrients in the diet. Furthermore, the new Mediterranean Diet was compatible with biodiversity and traditional and local food products.
- **Socialization.** The meal has an important cultural and social value. Eating around a table, as far as possible sitting down, and in the company of family and friends, are cultural traditions that should be preserved.

- Finally, **regular practice of physical activity** (at least 30 min every day) is needed to maintain a healthy body weight and balance energy intake.

During the past years, within the international debate on sustainability, food security, and nutrition, sustainable diets have emerged as a public health nutrition challenge as well as a critical issue for sustainable food systems. Sustainable diets are those diets with low environmental impacts, which contribute to food and nutrition security and to healthy life for present and future generations. Sustainable diets are protective and respectful of biodiversity and ecosystems, culturally acceptable, accessible, economically fair, and affordable; nutritionally adequate, safe, and healthy, while optimizing natural and human resources [20]. In this sense, the notion of the Mediterranean Diet has undergone a progressive evolution over the past 50 years, from a healthy dietary pattern to a model of sustainable diet.

The International Mediterranean Diet Foundation was founded in November 2014 as a center of multidisciplinary knowledge and expertise, internationally in order to revalorize the Mediterranean Diet. It works together with other organizations and experts in public health nutrition, food sciences, social anthropology, sociology, home economics, agriculture, environment, and cultural heritage.

In order to balance the worldwide interest in the Mediterranean Diet with increasing sustainability, especially environmental concerns, the International Mediterranean Diet Foundation updated the representation of the Mediterranean Diet Pyramid in order to provide a unified representation of the Mediterranean Diet as a sustainable dietary pattern encompassing the entirety of the Mediterranean Area. The main goal was to shift the perception of the Mediterranean diet benefits from a focus on the person—individual benefits—to a focus on the earth benefits for the planet as well as its populations (Fig. 1.4).

The First World Conference on the Mediterranean Diet held in Milan in 2016, reached a consensus on how to assess the adherence and the sustainability of the Mediterranean Diet at the country level; and how to reconstruct, at least partly, a sustainable eating culture and lifestyle more suited to the times and for all Mediterranean people by considering the diversity of their food cultures and food systems.

1.4 NUTRITIONAL COMPOSITION AND NUTRITIONAL ADEQUACY OF THE MEDITERRANEAN DIET

Nutritional adequacy is defined as the sufficient intake of essential nutrients, which are needed to fulfill nutritional requirements for optimal health. This term is also the comparison between the nutrient requirement and the intake of a specific population or individual.

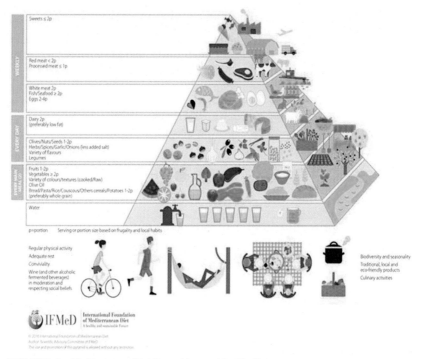

FIGURE 1.4   Mediterranean Diet Pyramid: a sustainable dietary pattern.

Over the last few decades the Mediterranean Diet has been recognized for its nutritional composition and effects on human health [5,18,22]. In fact, this dietary pattern has been considered to be one of the healthiest [22].

Several previous studies have shown the excellent profile of macronutrients and essential micronutrients of this dietary pattern [23,24]. The main mechanism, which is considered to explain this optimal nutritional quality, is the high consumption of food groups rich in minerals, vitamins, and antioxidants, such as vegetables, fruits, olive oil, and nuts.

Although previous evidence showed considerable variation in quantity of Mediterranean Diet components according to the region or country, there are several reasons why the traditional Mediterranean diet is a pattern with high nutritional quality and high adequacy of nutrient intake.

- This dietary pattern used to be a balanced diet and sufficiently caloric, with relatively low levels of energy-dense or nutrient-poor foods, such as processed meat, processed meals and fast food, but conversely having high levels of nutritionally dense foods.
- Although total fat intakes are not very uniform between Mediterranean countries, from <25% to >35% of energy from one area to another [6], in general the Mediterranean Diet contains a high percentage of daily

energy from total fat and has a healthy profile with regard to the quality of fat intake.

- Consequently, this diet has a high monounsaturated fat intake, a low saturated fat intake ($\leq 7-8$ %) and a moderate intake of polyunsaturated fat and long-chain omega-3 fatty acids (from fish and sea food).
- Furthermore, the ratio of monounsaturated/saturated fats is much higher in the Mediterranean basin than in other regions of the world, including North Europe and North America.
- A higher adherence to the Mediterranean Diet has been associated with a lower probability of not fulfilling the nutrient recommendation [18,25]. For example, in a recent study in the SUN (Seguimiento Universidad de Navarra in Spanish) cohort, participants who had a high adherence to a Mediterranean Diet were found to be at a lower risk of having micronutrient intake inadequacy [25]. Therefore, while the average number of nutrients with intakes lower than the recommended amount among participants with lower Mediterranean Diet score was 4.4, among those who had a higher score, the average number was 2.4 [26]. Similar results were observed in the PREDIMED (Prevención con Dieta Mediterránea in Spanish) trial, where this diet was also positively associated with overall micronutrient adequacy in elderly participants [26]. The results of other studies in other adult population and children have been very similar [21].
- Other characteristics of the traditional Mediterranean Diet are the high consumption of dietary fiber (from vegetables/plants), a low glycemic index, and glycemic load [14].
- This diet contains a high and varied intake of phenolic compounds, including flavonoids and various nonflavonoids such as phenolic acids, stilbenes, and lignans [4].
- This dietary pattern is rich in high antioxidant intake and total daily antioxidant capacity [5]. For example, one of the most typical foods of the Mediterranean dietary pattern, "extra-virgin" olive oil contains antioxidant and antiinflammatory compounds including tocopherols.
- The general pattern of alcohol consumption is quite different as compared to non-Mediterranean countries due to the type of alcohol. In the Mediterranean basin, the type of alcohol mainly consumed is wine, which is rich in antioxidants and antiinflammatory chemicals.

Finally, a recently published review has found that several aspects of food processing pertinent to the Mediterranean Diet could have an effect on the nutritional quality of foods, and these effects may also impact on health outcomes, especially in relation to oxidative stress and inflammation [27].

Because all these food groups or nutrient intakes could have a synergistic and antagonistic effect on health outcome, the study of overall dietary patterns and not just single nutrients has been suggested [28].

1.5 HOW TO ASSESS THE ADHERENCE TO THE MEDITERRANEAN DIET

The beneficial role of Mediterranean Diet in various health outcomes has been shown over the last decades in epidemiological studies conducted within Mediterranean, European, as well as in US populations. In this sense, the nutritional characteristics of the traditional Mediterranean Diet have been expressed through indexes or scales that quantify the adherence to this dietary pattern. However, to decide which of these numerous indexes is optimal is rather difficult because these indices/scores differ with respect to the level of detail of foods/nutrients included (e.g., meat overall, or, red meat and poultry separated), cut-offs used (i.e., medians, prespecified values, etc.), and range of values upon which individuals are ranked according to their adherence to overall Mediterranean Diet [29]. Moreover, most of those have been conducted within specific populations and they naturally express better the particularities of Mediterranean Diet in these populations. A related issue is whether indices developed for dietary data collected and quantified with specific methods/tools are appropriate for data collected/ quantified differently [29].

Adherence to the Mediterranean Diet has been widely appraised according to the score proposed by Trichopoulou et al. [30]. **The Mediterranean Diet Score** (MDS) includes nine components: vegetables, legumes, fruits and nuts, cereals, fish and seafood, meat and meat products, dairy products, moderate alcohol intake, and the ratio of monounsaturated fatty acids to saturated fatty acids. One point is assigned to persons whose consumption is at or above the sex-specific median of six components in agreement with the traditional Mediterranean Diet (vegetables, fruits/nuts, legumes, fish/seafood, cereals, and monounsaturated to saturated fatty acids ratio). The participant receives also 1 point if her or his intake is below the median for the two components not in line with the traditional Mediterranean Diet (meat or meat products and dairy products). For ethanol, 1 point is assigned only for moderate amounts of intake (5−25 g/d for women or 10−50 g/d for men). Therefore, this score can range from the highest possible (9 points reflecting maximum adherence) to the minimum possible (0 points reflecting no adherence at all).

The alternate Mediterranean Diet (aMED) score was developed by the Department of Nutrition of the Harvard T.H. Chan School of Public Health and is based on the Mediterranean Diet score by Trichopoulou et al. [31,32]. The original scale was modified by excluding potato products from the vegetable group, separating fruits, and nuts into two groups, eliminating the dairy group as a component of the score, including whole-grain products only as cereal sources, including only red and processed meats for the meat group, and assigning 1 point for alcohol intake between 5 and 15 g/d for women and 10 and 15 g/d for men.

Thus, the score rewards 1 point if intake is above the sample-specific median for vegetables, legumes, fruit, nuts, whole-grain cereals, fish, and the ratio monounsaturated to saturated fatty acids and 1 point for intake below the sample-specific median for red and processed meats. In addition, alcohol intake of 5−15 g/d for women and 10−15 g/d for men receives 1 point. A higher score represents a higher adherence to the Mediterranean Diet, with a score range between 0 and 9, similarly to the MDS.

The 14-Point Mediterranean Diet Adherence Screener (MEDAS) was developed and used in the PREDIMED trial, a nutrition intervention trial for primary prevention of cardiovascular disease. This screener consists of 12 questions on food consumption frequency and 2 questions on food intake habits considered characteristic of the Spanish Mediterranean Diet. Each question is scored 0 or 1. One point is given for using olive oil as the principal source of fat for cooking, preferring white meat over red meat, or for consuming: (1) 4 or more tablespoons (1 tablespoon = 13.5 g) of olive oil/d (including that used in frying, salads, meals eaten away from home, etc.); (2) 2 or more servings of vegetables/d; (3) 3 or more pieces of fruit/d; (4) <1 serving of red meat or sausages/d; (5) <1 serving of animal fat/d; (6) <1 cup (1 cup = 100 mL) of sugar-sweetened beverages/d; (7) 7 or more servings of red wine/wk; (8) 3 or more servings of pulses/wk; (9) 3 or more servings of fish/wk; (10) fewer than 2 commercial pastries/wk; (11) 3 or more servings of nuts/wk; or (12) 2 or more servings/wk of a dish with a traditional sauce of tomatoes, garlic, onion, or leeks sautéed in olive oil. If the condition is not met, 0 points are recorded for the category. The final PREDIMED score ranges from 0 to 14 [33].

The Mediterranean Diet 55 Score (MD55) was developed by Panagiotakos et al. within the Greek-ATTICA study [34]. Eleven main components of the Mediterranean Diet (nonrefined cereals, fruits, vegetables, potatoes, legumes, olive oil, fish, red meat, poultry, full fat dairy products, and alcohol) are used to construct this dietary pattern. For the consumption of items presumed to be close to this pattern scores of 0, 1, 2, 3, 4, and 5 are assigned according to the participants' frequency of consumption (no consumption, rare, frequent, very frequent, weekly and daily, respectively). For the consumption of foods presumed to be away from this pattern, the scores are assigned on a reverse scale. Especially for alcohol, score 5 for consumption of less than 300 ml/d, score 0 for consumption of more than 700 ml/d or none were assigned and scores 1−4 for consumption of 300−400, 400−500, 500−600, and 600−700 ml/d (100 ml = 12 g ethanol), respectively. Then a total score ranging from 0 to 55 was calculated.

The Mediterranean Adequacy Index (MAI) proposed by Alberti-Fidanza et al. [35] is obtained by dividing the sum of the percentage of total energy from typical Mediterranean food groups by the sum of the percentage of total energy from nontypical Mediterranean food groups.

Typical Mediterranean foods are bread, cereals, legumes, potatoes, vegetables, fruit, fish, red wine, and vegetable oils. Nontypical Mediterranean foods include milk, cheese, meat, eggs, animal fats and margarines, sweet beverages, cakes, pies and cookies, and sugar.

1.6 THE 2015 DIETARY GUIDELINES FOR AMERICANS AND THE HEALTHY DIETARY PATTERNS SUCH AS THE MEDITERRANEAN DIET

According to the Dietary Guidelines for Americans *"preventable, diet and lifestyle-related chronic diseases, including obesity, high blood pressure, type 2 diabetes, and cardiovascular disease, contribute to the high and rising costs of U.S. health care"*. According to these guidelines: *"the alarming current situation is that 65 percent of adult American females and 70 percent of adult American males are overweight or obese, and rates are highest in adults ages 40 years and older. Moreover, adults with overweight or obesity frequently have co-morbid conditions and higher chronic disease risk profiles that contribute substantially to higher health care costs. In fact, rates of elevated blood pressure, adverse blood lipid profiles..., and diabetes are highest in adults with elevated abdominal obesity (waist circumference greater than 102 cm in men, greater than 88 cm in women)"...* *Furthermore, many adults have personal health profiles in which multiple metabolic risk factors coexist and substantially increase risks for coronary heart disease, hypertension and stroke, diabetes, and other obesity-related co-morbidities* [36].

The Dietary Guidelines for Americans were first released in 1980, and since that time, they have provided science-based advice on promoting health and reducing risk of major chronic diseases through a healthy diet and regular physical activity. The Dietary Guidelines for Americans is published by the Federal government every 5 years. The 2015 Dietary Guidelines Advisory Committee was established for the single, time-limited task of reviewing the 2010 edition of Dietary Guidelines for Americans and developing nutrition and related health recommendations to the Federal government for its subsequent development of the 2015 edition.

In this sense, the 2015 Dietary Guidelines Advisory Committee advocates *"achieving healthy dietary patterns through healthy food and beverage choices rather than with nutrient or dietary supplements except as needed."* Healthy eating patterns can be achieved for a variety of eating styles, including the "Healthy U.S.-Style Pattern," the "Healthy Vegetarian-Style Pattern," and the "Healthy Mediterranean-Style Pattern." According to this committee strong and consistent evidence demonstrates that dietary patterns associated with decreased risk of cardiovascular disease are characterized by higher consumption of vegetables, fruits, whole grains, low-fat dairy, and seafood, and lower consumption of red and processed meat, and lower intakes of refined grains, and sugar-sweetened foods and beverages relative to less healthy patterns. Regular consumption of nuts and legumes and moderate consumption of alcohol also are shown to be components of a beneficial dietary pattern in most studies. To reach this conclusion the Dietary Guidelines Advisory Committee examined research compiled in the NEL (USDA's Nutrition Evidence Library) Dietary Patterns Systematic Review Project, which included 55 articles summarizing evidence from 52 prospective cohort studies and 7 randomized clinical trials, and other guidelines and reports. The Committee drew additional evidence and effect size estimates from six published systematic reviews/meta-analyses published since 2008 that included one or more studies not covered in the NEL or other reports. In total, 142 articles were considered in these reports. Some of these results are derived from systematic reviews and clinical trials designed to test the beneficial role of the Mediterranean Diet. Therefore, at this point it is important to describe the epidemiological evidences that have supported the beneficial role of the Mediterranean Diet and that have made to include it in the most recent American Dietary Guidelines as a healthy dietary pattern in the prevention of cardiovascular disease and other cardio-metabolic conditions such as obesity, diabetes, or hypertension.

1.7 EPIDEMIOLOGICAL EVIDENCES REGARDING THE BENEFICIAL ROLE OF THE MEDITERRANEAN DIET IN CARDIOVASCULAR DISEASE AND CARDIOVASCULAR RISK FACTORS

Since its origins, when Ancel Keys initiated his studies on the Mediterranean Diet, the principal disease outcome analyzed with respect to its relation with the Mediterranean Diet has been the cardiovascular disease, and particularly coronary heart disease.

During the second half of the Twentieth century, most of the research done was oriented to cardiovascular disease risk factors and only at the end of the century were large observational cohorts conducted to increase the evidence regarding the relation between the adherence to the Mediterranean Diet and risk of cardiovascular disease. In fact, an enormous growth in the amount and quality of available scientific evidence during the last 25 years supports that the Mediterranean Diet might be one of the healthiest dietary patterns in the world and could be the most plausible explanation for the increased longevity and lower cardiovascular disease rates observed in the Mediterranean countries.

Some of these studies have been carried out, as expected, within Mediterranean populations. It is the case of the SUN study, a Spanish prospective cohort study based on university graduates that followed-up 3609 participants (initially free of cardiovascular disease) during 4.9 years. Participants with the highest adherence to the Mediterranean Diet (score > 6 according to the MDS from Trichoupolou) exhibited a lower cardiovascular risk (significant risk reduction = 59%) compared to those with the lowest score (<3). For each 2-point increment in the score, the adjusted relative reduction in risk was 20% for total cardiovascular disease and 26% for coronary heart disease [37]. These researchers found similar results in a subsequent analysis using the MEDAS questionnaire to assess the adherence to the Mediterranean dietary pattern [38].

Within Spain, it is also remarkable the results obtained from the EPIC-Spain. The EPIC study (European Prospective Investigation into Cancer and Nutrition) is one of the largest cohort studies in the world, with more than half a million participants recruited across 10 European countries and followed for almost 15 years. This study examined the relation between Mediterranean Diet adherence and risk of incident coronary heart disease events in the five Spanish centers of the EPIC study. After following up to 41,078 participants for a mean of 10.4 years, the researchers found that high compared with low relative Mediterranean Diet adherence was associated with a significant reduction in coronary heart disease risk (relative risk reduction = 40%, 95% confidence interval = 23%−53%). A 1-unit increase in relative Mediterranean Diet score was associated with a significant reduction in the risk (6%) [39].

The EPIC-Greece has also provided interesting results regarding the association between the adherence to the Mediterranean Diet and cardiovascular risk. Dilis et al. found a significant relative reduction in coronary mortality (25% for women and 19% for men) associated to a 2-point increase in the MDS. The association of adherence to the Mediterranean Diet with coronary heart disease incidence was again inverse, but weaker and no significant [40]. Misirli et al. also found that increased adherence to the Mediterranean Diet, as measured by 2-point increments in score, was inversely associated with cerebrovascular incidence and mortality in this sample. These inverse trends were mostly evident among women and with respect to ischemic rather than hemorrhagic cerebrovascular disease [41].

All these effects have also been observed in other European cohort studies outside the Mediterranean such as the EPIC-Netherland [42], the Doetinchem Cohort Study [43], the Health Alcohol and Psychosocial factors in Eastern Europe (HAPIEE) study [44,45], the Swedish Mammography Cohort [46], or the Cohort of Swedish Men [47].

Within the US, it is worthy highlighting the results obtained in the Nurses' Health Study (NHS) from the University of Harvard. In this prospective cohort study composed by more than 100,000 registered female nurses, participants with the highest adherence to the aMED showed the lowest risk for both coronary heart disease and stroke when were compared with those with the lowest adherence. Cardiovascular disease mortality was significantly lower among women in the top quintile of the aMED (relative risk reduction = 39%; 95% confidence interval = 24%−51%; with a significant dose−response relationship) [31]. More recently, these results have been confirmed in a more complete analysis from the same institution including data not only from the NHS but also from the Health Professional Follow-up Study (HSPH). The HSPH is, likewise, a cohort prospective study but, in this case, based on males [32].

Also Bertoia et al. in the Women's Health Initiative study and Gardener et al. in the Northern Manhattan Study have found a protective role of the Mediterranean Diet in the risk of sudden cardiac death [48] and of a composite outcome of ischemic stroke, myocardial infarction, or vascular death [49], respectively.

Most of the results from these observational prospective studies have been summarized into several systematic reviews and meta-analyses. In the meta-analysis of Sofi et al. which included cohort studies published up to June 2013 and performed in an overall population of 4,172,412 subjects, 2-point increase in the MDS was related with a significant 8 % reduction of overall mortality and a 10% reduced risk of cardiovascular disease [50]. Other recent systematic review has confirmed the reported results [51].

One of the main limitations of observational studies is the lack of control of several potential confounders that might bias the results. Hence, the potential effect of the Mediterranean Diet on cardiovascular disease could be in part explained by the cooccurrence of other lifestyle-related factors such as physical activity, alcohol intake, or smoking. Thus, one of the most important aspects in observational epidemiology is to obtain an adequate control of these possible confounders. When the lack of or inadequate control for some of these potential confounders and the presence of residual confounding exist, the interpretation of the findings obtained from observational studies demands caution. Consequently, beyond prospective cohort studies, a few clinical or community trials with an experimental design that results more rigorous and methodologically stronger than observational studies have been carried out to assess the effect of the nutritional intervention with Mediterranean Diet in the primary and secondary prevention of cardiovascular disease. These studies have exponentially increased the level and the quality of the evidence around the Mediterranean Diet and cardiovascular disease in the last decades.

The Lyon Diet Heart Study was a randomized secondary prevention trial aimed at testing whether a Mediterranean-type diet may reduce the rate of recurrence after a first myocardial infarction. The trial found that subjects following the Mediterranean-Style diet had a 50%−70% lower risk of recurrent heart disease, as measured by three different combinations of outcome measures including (a) cardiac death and nonfatal heart attacks; (b) the preceding plus unstable angina, stroke, heart failure, and pulmonary or peripheral embolism; and (c) all of these measures plus events that required hospitalization. However, as we have already mentioned, this trial was a secondary prevention trial only including survivors from a myocardial infarction, the number of events was modest, and no special consideration was given to olive oil, which is the main source of fat in Mediterranean countries [52].

In this sense, the PREDIMED trial is a parallel-group, multicenter, randomized trial designed to test the efficacy of two Mediterranean diets (one supplemented with extra-virgin olive oil and another with nuts), as compared with a control diet (advice on a low-fat diet), on primary cardiovascular prevention in high-risk patients. This trial (with 7447 participants) is the largest Mediterranean Diet trial carried out until now. In its main analysis, the PREDIMED trial found that an energy-unrestricted Mediterranean Diet supplemented with either extra-virgin olive oil or nuts had favorable effects in high-risk participants compared to the control group who were advised to reduce dietary fat intake. An approximately 30% decrease in risk of major cardiovascular events (a composite endpoint including myocardial infarction, stroke, and deaths) was observed resulting, thus, in an absolute risk reduction of approximately three major cardiovascular events per 1000 person-years [53].

This landmark trial has been extensively referenced in the new Dietary Guidelines for Americans and is in part responsible for the inclusion of a Healthy Mediterranean-Style Pattern as one of the possible healthy eating patterns proposed by Dietary Guidelines Advisory Committee to be adopted by the American population.

At the present time, other trials have been designed to ascertain the role of the Mediterranean Diet (especially energy-restricted diet) in cardiovascular disease although none scientific result has been already published.

The CARDIVEG study is a crossover clinical trial that will test whether there is a difference between a vegetarian calorie-restricted diet and a Mediterranean calorie-restricted diet in reducing total weight and ameliorating the cardiovascular risk profile of a clinically healthy group of subjects [54]. The PREDIMED-PLUS is the continuation of the PREDIMED trial and has been designed to demonstrate that the adherence to an energy-restricted Mediterranean Diet associated with daily physical activity as compared to the adherence to an energy-unrestricted Mediterranean Diet is able to improve even more the cardiovascular risk profile and to achieve a maintained long-term weight loss.

TABLE 1.1 Main Results of the PREDIMED Trial

Reference	Sample	Follow-Up (y)	Outcome	Results-Conclusions
Clinical Events				
Martínez-González et al. Circulation 2014; 130:18–26	6705 participants	4.7	Atrial fibrillation	As compared to the control group, the authors found a risk reduction of 38% for the MedDiet + EVOO group. No effect was found for the MedDiet + nuts group
Salas-Salvadó et al. Ann Intern Med 2014; 160:1–10	Overall sample	4.1	Type 2 diabetes	As compared to the control group, the authors found a risk reduction of 40% for the MedDiet + EVOO group and of 30% for both MedDiets
Ruiz-Canela et al. JAMA 2014; 311:415–17	7465 participants	4.8	Peripheral artery disease	As compared to the control group, the authors found a risk reduction = 64% for MedDiet + EVOO group and of 50% for the MedDiet + nuts group
Babio et al. CMAJ 2014;186(17):E649–57	5801 participants	4.8	Metabolic syndrome (incidence and reversion)	MedDiet + EVOO or MedDiet + nuts were not associated with the onset of metabolic syndrome, but such diets are more likely to cause reversion of the condition
Estruch et al. N Engl J Med 2013; 368(14): 1279–90	Overall sample	4.8	Cardiovascular disease	As compared to the control group, the authors found a risk reduction = 70% for both MedDiet + EVOO and MedDiet + nuts groups
Sánchez-Villegas et al. BMC Med 2013;11:208	Overall sample	5.4	Depression incidence	MedDiet + nuts was associated with a reduced risk of depression among diabetics
Salas-Salvadó et al. Arch Intern Med 2008;168(22):2449–58	1224 participants	1	Metabolic syndrome (1-year prevalence and incidence)	A statistical difference in 1-year prevalence of metabolic syndrome was found for MedDiet + nuts (as compared to the control group). Incident rates of metabolic syndrome were not significantly different among groups

Changes in Markers and Risk Factors for Cardiovascular Disease

Reference	Sample		Outcome	Results
Estruch et al. Lancet Diabetes Endocrinol 2016;4(8):666−76	Overall sample	5	Bodyweight and waist circumference	As compared to the control group, statistical significant difference in changes in bodyweight in the MedDiet + EVOO group. Statistical significant differences in changes in waist circumference for both MedDiets
				Mediterranean Diet was associated with decreases in bodyweight and less gain in central adiposity compared with a control diet
Stomiolo et al. Eur J Nutr 2017;56(1):89−97	Nonsmoking women with hypertension (90 participants)	1	Blood pressure, serum nitric oxide, and endothelin-1 and related gene expression	A negative correlation was observed between changes in nitric oxide metabolites concentration and blood pressure after the intervention with MedDiet + EVOO. Systolic blood pressure reduction was related to an impairment of serum ET-1 concentrations after the intervention with MedDiet + nuts
Casas et al. PLoS One 2014;9(6):e100084	164 participants	1	Classical cardiovascular risk factors and inflammatory biomarkers of atherosclerosis and plaque vulnerability	Adherence to the MedDiet was associated with a decrease in systolic and diastolic blood pressure, LDL-cholesterol and inflammatory biomarkers related to plaque instability, and with an increase in serum markers of atheroma plaque stability
Fitó et al. Eur J Heart Fail 2014;16 (5):543−50	930 participants	1	Heart failure biomarkers	MedDiet + EVOO was associated with a decrease in in vivo oxidized low-density lipoprotein, lipoprotein(a) plasma concentrations, and N-terminal probrain natriuretic peptide
Damasceno et al. Atherosclerosis 2013;230(2):347−53	169 participants	1	Lipoprotein subclasses (particle concentrations and size)	Lipoprotein subfractions are shifted to a less atherogenic pattern by consumption of MedDiet + nuts

(Continued)

TABLE 1.1 (Continued)

Reference	Sample	Follow-Up (y)	Outcome	Results-Conclusions
Toledo et al. BMC Med 2013;11:207	Overall sample	4	Blood pressure	Both MedDiet and low-fat diet exerted beneficial effects on blood pressure but lower values of diastolic blood pressure were found for MedDiet + EVOO or MedDiet + nuts
Zamora-Ros et al. Nutr Metab Cardiovasc Dis 2013;23(12):1167−74	564 participants	1	Nonenzymatic antioxidant capacity	Plasma nonenzymatic antioxidant capacity increased in MedDiet + EVOO and MedDiet + nuts groups after 1 year of intervention
Razquin et al. Eur J Clin Nutr 2009; 63(12):1387−93	187 participants	3	Plasma antioxidant capacity and body weight gain	MedDiet, especially MedDiet + EVOO, was associated with higher levels of plasma total antioxidant capacity (TAC). Plasma TAC was related to a reduction in body weight
Fitó et al. Arch Intern Med 2007;167 (11):1195−203	372 participants	3 months	Lipoprotein oxidation	Reduction of oxidized low-density lipoprotein (LDL) levels in MedDiet + EVOO and MedDiet + nuts groups with a significant difference for MedDiet + EVOO as compared to the control group
Estruch et al. Ann Intern Med 2006;145 (1):1−11	772 participants	3 months	Body weight, blood pressure, lipid profile, glucose levels, and inflammatory molecules	Compared with the low-fat diet, both MedDiets produced beneficial changes in glucose and insulin levels, systolic and diastolic blood pressure, HOMA index, HDL-Cholesterol, and Cholesterol-HDL Cholesterol ratio

MedDiet, Mediterranean Diet; *EVOO*, Extra-virgin olive oil.

Moreover, the Mediterranean Diet has been also associated with a reduced risk of metabolic syndrome, diabetes, hypertension, or total mortality among subjects with cardiovascular risk factors and with an improvement in endothelium vasodilatation, lipoprotein levels, and glycemic and weight control or with a reduction in low-grade inflammation. Moreover, this diet shows important antioxidant properties.

Although some evidences proceed from Mediterranean and non-Mediterranean prospective cohorts such as the mentioned SUN, EPIC, NHS, or HPFS, again the findings reported for the PREDIMED trial deserve special consideration. In fact, this trial showed significant improvements in classical and emerging cardiovascular risk factors (for example, blood pressure, insulin sensitivity, lipid profiles, lipoprotein particles, inflammation, oxidative stress, and carotid atherosclerosis). Moreover, according to this trial, the effects of the Mediterranean Diet might differ depending on genetic variants. Several gene−diet interactions in determining both intermediate and cardiovascular phenotypes were found. Thus, beneficial effects of the intervention with Mediterranean Diet supplemented with extra-virgin olive oil or nuts showed interactions with several genetic variants (TCF7L2, APOA2, MLXIPL, LPL, FTO, M4CR,COX-2, GCKR and SERPINE1).

Table 1.1 shows some of the most important results published regarding the role of the Mediterranean Diet in several cardiovascular risk factors in the PREDIMED trial.

1.8   CONCLUSIONS

The healthy-eating and antique model called "Mediterranean Diet" has been recently recommended by the 2015 Dietary Guidelines Advisory Committee as a recommendable dietary pattern to prevent cardiovascular disease. One landmark trial named PREDIMED that has demonstrated its beneficial effect on several cardiovascular risk factors as well as on cardiovascular disease itself has been, in part responsible, to add this pattern in the most recent American Dietary Guidelines. Throughout this book, the reader will know all the biological and epidemiological evidences that support this important role of the Mediterranean Diet and of each of its components such as olive oil, cereals, legumes, nuts, fish, or vegetables and fruits.

REFERENCES

[1] Simopoulos AP, Visioli F, Koletzko B. Mediterranean Diets. World Review of Nutrition and Dietetics, 87. Switzerland: S. Karger AG; 2000.
[2] Keys A, Menotti A, Karvonen MJ, Aravanis C, Blackburn H, Buzina R, et al. The diet and 15-year death rate in the seven countries study. Am J Epidemiol 1986;124:903−15.
[3] Menotti A, Puddu PE. How the Seven Countries Study contributed to the definition and development of the Mediterranean diet concept: a 50-year journey. Nutr Metab Cardiovasc Dis 2015;25(3):245−52.

[4] Gerber M, Hoffman R. The Mediterranean diet: health, science and society. Br J Nutr 2015;113(2):S4−10.

[5] Davis C, Bryan J, Hodgson J, Murphy K. Definition of the Mediterranean Diet; a Literature Review. Nutrients 2015;7(11):9139−53.

[6] Willett WC, Sacks F, Trichopoulou A, Drescher G, Ferro-Luzzi A, Helsing E, et al. Mediterranean diet pyramid: a cultural model for healthy eating. Am J Clin Nutr 1995;61 (Suppl):S1402−6.

[7] Trichopoulou A, Lagiou P. Healthy traditional Mediterranean diet: an expression of culture, history, and lifestyle. Nutr Rev 1997;55(11 Pt 1):383−9.

[8] Donini LM, Serra-Majem L, Bulló M, Gil Á, Salas-Salvadó J. The Mediterranean diet: culture, health and science. Br J Nutr 2015;113(2)):S1−3.

[9] Sofi F. The Mediterranean diet revisited: evidence of its effectiveness grows. Curr Opin Cardiol 2009;24(5):442−6.

[10] Matalas AL, Zampelas A, Stavrinos V, Wolinsky I. The Mediterranean Diet: constituents and health promotion. Florida: CRC Press; 2000.

[11] Renna M, Rinaldi VA, Gonnella M. The Mediterranean Diet between traditional foods and human health: the culinary example of Puglia (Southern Italy). Int J Gastron Food Sci 2015;2:63−71.

[12] Reguant-Aleix J, Arbore MR, Bach-Faig A, Serra-Majem L. Mediterranean heritage: an intangible cultural heritage. Public Health Nutr 2009;12(9A):1591−4.

[13] UNESCO. Representative list of the intangible cultural heritage of humanity. http://www. unesco.org/culture/ich/en/RL/00394. 2010.

[14] Trichopoulou A, Martínez-González MA, Tong TY, Forouhi NG, Khandelwal S, Prabhakaran D, et al. Definitions and potential health benefits of the Mediterranean diet: views from experts around the world. BMC Med 2014;12:112.

[15] Da Silva R, Bach-Faig A, Raidó Quintana B, Buckland G, Vaz de Almeida MD, Serra-Majem L. Worldwide variation of adherence to the Mediterranean diet, in 1961-1965 and 2000-2003. Public Health Nutr 2009;12:1676−84.

[16] Serra-Majem L, Trichopoulou A, Ngo de la Cruz J, Cervera P, García Alvarez A, La Vecchia C, et al. International Task Force on the Mediterranean Diet. Does the definition of the Mediterranean diet need to be updated? Public Health Nutr 2004;7(7):927−9.

[17] Medina F. Mediterranean diet, culture and heritage: challenges for a new conception. Public Health Nutr 2009;12(9A):1618−20.

[18] Oldwayspt. Inspiring Good Health Through Cultural Food Traditions. http://oldwayspt. org/history-mediterranean-diet-pyramid.

[19] Bach-Faig A, Berry EM, Lairon D, Reguant J, Trichopoulou A, Dernini S, et al. Mediterranean Diet Foundation Expert Group. Mediterranean diet pyramid today. Science and cultural updates. Public Health Nutr 2011;14(12A):2274−84.

[20] Food and Agriculture Organization of the United Nations. http://www.fao.org/docrep/016/ i3004e/i3004e04.pdf.

[21] Castro-Quezada I, Román-Viñas B, Serra-Majem L. The Mediterranean diet and nutritional adequacy: a review. Nutrients 2014;6(1):231−48.

[22] Serra-Majem L, Bach-Faig A, Raidó-Quintana B. Nutritional and cultural aspects of the Mediterranean diet. Int J Vitam Nutr Res 2012;82(3):157−62.

[23] Maillot M, Issa C, Vieux F, Lairon D, Darmon N. The shortest way to reach nutritional goals is to adopt Mediterranean food choices: evidence from computer-generated personalized diets. Am J Clin Nutr 2011;94:1127−37.

[24] Serra-Majem L, Bes-Rastrollo M, Román-Viñas B, Pfrimer K, Sánchez-Villegas A, Martínez-González MA. Dietary patterns and nutritional adequacy in a Mediterranean country. Br J Nutr 2009;101(2):S21–8.

[25] Zazpe I, Sánchez-Taínta A, Santiago S, de la Fuente-Arrillaga C, Bes-Rastrollo M, Martínez JA, et al. Association between dietary carbohydrate intake quality and micronutrient intake adequacy in a Mediterranean cohort: the SUN (Seguimiento Universidad de Navarra) Project. Br J Nutr 2014;14;111(11):2000–9.

[26] Sánchez-Tainta A, Zazpe I, Bes-Rastrollo M, Salas-Salvadó J, Bullo M, Sorlí JV, et al. PREDIMED study investigators. Nutritional adequacy according to carbohydrates and fat quality. Eur J Nutr 2016;55(1):93–106.

[27] Hoffman R, Gerber M. Food processing and the Mediterranean Diet. Nutrients 2015; 7(9):7925–64.

[28] Preedy V, Watson RR. The Mediterranean diet: an evidence-based approach. New York: Elsevier; 2015.

[29] Bamia C, Martimianaki G, Trichopoulou A. Assessment of adherence to the Mediterranean diet with different indices. Revitalizing the Mediterranean Diet. 1st World Conference on the Mediterranean Diet. Book of Abstracts; 2016.

[30] Trichopoulou A, Costacou T, Bamia C, Trichopoulos D. Adherence to a Mediterranean diet and survival in a Greek population. N Engl J Med 2003;348(26):2599–608.

[31] Fung TT, Rexrode KM, Mantzoros CS, Manson JE, Willett WC, Hu FB. Mediterranean diet and incidence of and mortality from coronary heart disease and stroke in women. Circulation 2009;119(8):1093–100.

[32] Sotos-Prieto M, Bhupathiraju SN, Mattei J, Fung TT, Li Y, Pan A, et al. Changes in diet quality scores and risk of cardiovascular disease among US men and women. Circulation 2015;132(23):2212–19.

[33] Schröder H, Fitó M, Estruch R, Martínez-González MA, Corella D, Salas-Salvadó J, et al. A short screener is valid for assessing Mediterranean diet adherence among older Spanish men and women. J Nutr 2011;141(6):1140–5.

[34] Panagiotakos DB, Pitsavos C, Stefanadis C. Dietary patterns: a Mediterranean diet score and its relation to clinical and biological markers of cardiovascular disease risk. Nutr Metab Cardiovasc Dis 2006;16(8):559–68.

[35] Alberti-Fidanza A, Fidanza F. Mediterranean Adequacy Index of Italian diets. Public Health Nutr 2004;7(7):937–41.

[36] Dietary Guidelines for Americans. https://health.gov/dietaryguidelines/2015/.

[37] Martínez-González MA, García-López M, Bes-Rastrollo M, Toledo E, Martínez-Lapiscina EH, Delgado-Rodriguez M, et al. Mediterranean diet and the incidence of cardiovascular disease: a Spanish cohort. Nutr Metab Cardiovasc Dis 2011;21(4):237–44.

[38] Domínguez LJ, Bes-Rastrollo M, de la Fuente-Arrillaga C, Toledo E, Beunza JJ, Barbagallo M, et al. Similar prediction of total mortality, diabetes incidence and cardiovascular events using relative- and absolute-component Mediterranean diet score: the SUN cohort. Nutr Metab Cardiovasc Dis 2013;23(5):451–8.

[39] Buckland G, González CA, Agudo A, Vilardell M, Berenguer A, Amiano P, et al. Adherence to the Mediterranean diet and risk of coronary heart disease in the Spanish EPIC Cohort Study. Am J Epidemiol 2009;170(12):1518–29.

[40] Dilis V, Katsoulis M, Lagiou P, Trichopoulos D, Naska A, Trichopoulou A. Mediterranean diet and CHD: the Greek European Prospective Investigation into Cancer and Nutrition cohort. Br J Nutr 2012;108(4):699–709.

[41] Misirli G, Benetou V, Lagiou P, Bamia C, Trichopoulos D, Trichopoulou A. Relation of the traditional Mediterranean diet to cerebrovascular disease in a Mediterranean population. Am J Epidemiol 2012;176(12):1185−92.

[42] Hoevenaar-Blom MP, Nooyens AC, Kromhout D, Spijkerman AM, Beulens JW, van der Schouw YT, et al. Mediterranean style diet and 12-year incidence of cardiovascular diseases: the EPIC-NL cohort study. PLoS One 2012;7(9):e45458.

[43] Hoevenaar-Blom MP, Spijkerman AM, Boshuizen HC, Boer JM, Kromhout D, Verschuren WM. Effect of using repeated measurements of a Mediterranean style diet on the strength of the association with cardiovascular disease during 12 years: the Doetinchem Cohort Study. Eur J Nutr 2014;53(5):1209−15.

[44] Stefler D, Malyutina S, Kubinova R, Pajak A, Peasey A, Pikhart H, et al. Mediterranean diet score and total and cardiovascular mortality in Eastern Europe: the HAPIEE study. Eur J Nutr 2015;56(1):421−9.

[45] Tsivgoulis G, Psaltopoulou T, Wadley VG, Alexandrov AV, Howard G, Unverzagt FW, et al. Adherence to a Mediterranean diet and prediction of incident stroke. Stroke 2015; 46(3):780−5.

[46] Tektonidis TG, Åkesson A, Gigante B, Wolk A, Larsson SC. A Mediterranean diet and risk of myocardial infarction, heart failure and stroke: a population-based cohort study. Atherosclerosis 2015;243(1):93−8.

[47] Tektonidis TG, Åkesson A, Gigante B, Wolk A, Larsson SC. Adherence to a Mediterranean diet is associated with reduced risk of heart failure in men. Eur J Heart Fail 2016;18(3):253−9.

[48] Bertoia ML, Triche EW, Michaud DS, Baylin A, Hogan JW, Neuhouser ML, et al. Mediterranean and Dietary Approaches to Stop Hypertension dietary patterns and risk of sudden cardiac death in postmenopausal women. Am J Clin Nutr 2014;99(2):344−51.

[49] Gardener H, Wright CB, Gu Y, Demmer RT, Boden-Albala B, Elkind MS, et al. Mediterranean-style diet and risk of ischemic stroke, myocardial infarction, and vascular death: the Northern Manhattan Study. Am J Clin Nutr 2011;94(6):1458−64.

[50] Sofi F, Macchi C, Abbate R, Gensini GF, Casini A. Mediterranean diet and health status: an updated meta-analysis and a proposal for a literature-based adherence score. Public Health Nutr 2014;17(12):2769−82.

[51] Martinez-Gonzalez MA, Bes-Rastrollo M. Dietary patterns, Mediterranean diet, and cardiovascular disease. Curr Opin Lipidol 2014;25(1):20−6.

[52] de Lorgeril M, Salen P, Martin JL, Monjaud I, Delaye J, Mamelle N. Final Report of the Lyon Diet Heart Study. Circulation 1999;99:779−85.

[53] Estruch R, Ros E, Salas-Salvadó J, Covas MI, Corella D, Arós F, et al. PREDIMED Study Investigators. Primary prevention of cardiovascular disease with a Mediterranean diet. N Engl J Med 2013;368(14):1279−90.

[54] Sofi F, Dinu M, Pagliai G, Cesari F, Marcucci R, Casini A. Mediterranean versus vegetarian diet for cardiovascular disease prevention (the CARDIVEG study): study protocol for a randomized controlled trial. Trials 2016;17(1):233.

Chapter 2

Epidemiological and Nutritional Methods

Estefanía Toledo

2.1   HOW DO WE CLASSIFY THE EXPOSURE IN NUTRITIONAL EPIDEMIOLOGY?

In order to assess if a certain exposure is associated to the outcome that we are interested in studying, we need to classify our participants according to the exposure. This is due to the fact that we need to compare the incidence of the disease among those who are exposed and those who are not exposed.

If the occurrence of the disease is higher among exposed than among non-exposed, we will consider the exposure as a risk factor for the outcome of interest. Contrarily, if the occurrence of the disease is lower among exposed than among non-exposed, we will say that the exposure that we are studying is a protective factor against the outcome that we are considering.

The question that arises at this point is: how do we classify our participants according to their exposure? Some exposures have an intuitive way of classification. For example, if we wanted to study the effects of smoking on health, we could classify the participants into three groups: never smokers, former smokers, and current smokers.

However, sometimes the groups are not so intuitive. For instance, let us consider the association between fiber intake and the incidence of cardiovascular disease. How can we classify our participants according to their fiber intake? We certainly need to have groups that comprise participants with low and with high fiber intake. But, where do we set our cut-off points? It is common practice in nutritional epidemiology to use quantiles. Quantiles are cut-off points that divide the sample into equally sized groups (Fig. 2.1). Accordingly, tertiles divide the sample into three equally sized groups. The first group will comprise those 33% participants with the lowest fiber consumption, the second one the 33% of the sample with a medium fiber intake, and the third one will comprise those participants with the highest fiber consumption.

The Prevention of Cardiovascular Disease Through the Mediterranean Diet.
DOI: http://dx.doi.org/10.1016/B978-0-12-811259-5.00002-0

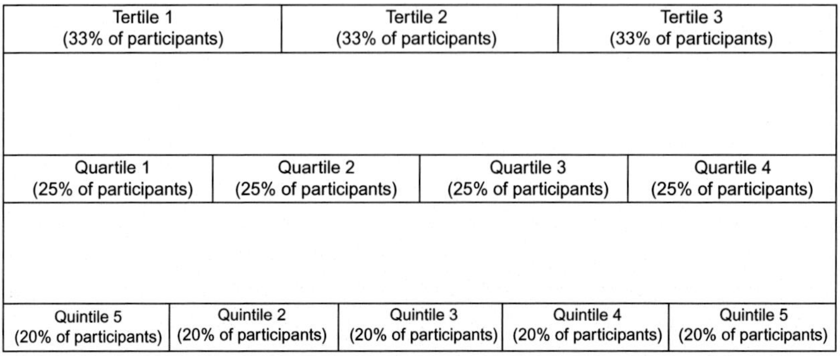

Tertile 1 (33% of participants)		Tertile 2 (33% of participants)		Tertile 3 (33% of participants)	
Quartile 1 (25% of participants)	Quartile 2 (25% of participants)	Quartile 3 (25% of participants)		Quartile 4 (25% of participants)	
Quintile 5 (20% of participants)	Quintile 2 (20% of participants)	Quintile 3 (20% of participants)	Quintile 4 (20% of participants)	Quintile 5 (20% of participants)	

FIGURE 2.1   Classification of participants according to quantiles.

Besides tertiles, the most widely used quantiles in nutritional epidemiology are quartiles, that divide the sample into four equally sized groups, and quintiles, that divide the sample into five equally sized groups.

2.2   HOW DO WE MEASURE THE OCCURRENCE OF DISEASE?

Quantifying the occurrence of the disease is a cornerstone in epidemiology because establishing associations is based on comparing the frequency of disease occurrence among exposed and among non-exposed participants.

Depending on the available information, we will be able to quantify occurrence of disease with different measures of frequency.

2.2.1   Prevalence Proportion

The easiest way of quantifying the disease is determining the proportion of people who have the disease at the present time [1]. The prevalence is thus calculated as the number of people who have the disease over the total number of observed people (Fig. 2.2).

2.2.2   Prevalence Odds

For the prevalence proportion, we considered in the denominator all participants, that is, those that have the disease and those that do not have the disease.

The prevalence odds, on the contrary, includes in the denominator only those subjects who do not have the disease [1]. Therefore, it expresses the number of people with the disease over the number of people without the disease.

$$\text{Prevalence proportion} = \frac{\text{\# of people with the disease}}{\text{total \# of people}}$$

$$\text{Prevalence odds} = \frac{\text{\# of people with the disease}}{\text{\# of people without the disease}}$$

$$\text{Cumulative incidence} = \frac{\text{\# of people who experience the disease during a certain period}}{\text{\# of people free of the disease at baseline}}$$

$$\text{Incidence rate} = \frac{\text{\# of people who experience the disease during a certain period}}{\text{Total person--time}}$$

FIGURE 2.2   Calculation of prevalence proportion, cumulative incidence, and incidence rate.

2.2.3   Cumulative Incidence

The incidence tries to capture information on new cases of the disease. It is calculated as "the number of new cases of a disease that occur during a specified period of time in a population at risk for developing the disease" [2] (Fig. 2.2). An incidence rate of 3% is only interpretable as long as the time frame of follow-up is explicit. The scenario is different if 3% of the participants sicken during 1 month of follow-up or if they do over a 10-year period. Therefore, the time frame to which the incidence rate refers to has always to be made explicit.

Also, the fact of being at risk for developing the disease has to be considered when it comes to calculate an incidence rate. This means, that only people who were free of the disease at study inception can be considered as susceptible of developing the disease. Thus, only people who were free of the disease at study baseline are going to be included in the denominator.

Finally, in order to calculate the cumulative incidence, participants need to be observed until the end of the follow-up period or the disease onset, whichever occurs first.

2.2.4   Incidence Rate or Incidence Density

Sometimes it is not feasible to achieve complete follow-up for all the study participants. This fact may be due to several reasons. For example, some participants may be lost to follow-up and other may die due to an outcome different from the one we are interested in studying. Considering the information of only those participants with complete follow-up would imply disregarding the information of those participants who were followed up for some time. A way of avoiding this last problem is to sum the time periods that each person has been followed up and calculate the total time of person-time at risk. This measure is the denominator of the incidence rate. The numerator is going to be the number of new cases during a given period of follow-up [2] (Fig. 2.2).

2.2.5 Hazard Function

The hazard function is the instantaneous probability of having an event at time t, that is, the instantaneous incidence rate given that someone has not developed the outcome of interest up to that time [3]. The hazard is thus not a single figure but a function that changes over time. It provides information not only about the disease occurrence, but also about the speed at which the disease occurs.

2.3 HOW DO WE ASSESS IF A CERTAIN EXPOSURE IS ASSOCIATED TO A CERTAIN OUTCOME?

As previously mentioned, we compare the frequency of the outcome among those that are exposed to the frequency of the outcome among those who are not exposed in order to assess if an exposure is associated to a certain outcome. This comparison is usually done in a relative way, that is, we divide the frequency of the outcome among exposed over the frequency of the outcome among non-exposed. Thus, we obtain ratios that are called measures of association. Depending on the measure of frequency that we use, we will calculate different measures of association.

2.3.1 Relative Risk

The relative risk is a measure of association that is usually used in cohort studies (see later). In this type of studies, the risk of developing the disease can be estimated among those who are exposed and among those who are not exposed. The relative risk is then defined as the risk among exposed over the risk among non-exposed (Fig. 2.3) [2].

When the incidence of the disease among exposed equals the incidence of the disease among non-exposed, the relative risk will have a value of 1. In this case, we would conclude that there is no association between the exposure and the outcome of interest.

When the risk among exposed is higher than the risk among non-exposed, the relative risk will have a value greater than 1. In this case, we will consider the exposure as a risk factor for the outcome that we are studying. For example, a relative risk of 1.5 means that the risk among exposed is 1.5 times the risk among non-exposed. Accordingly, this can be interpreted as the exposed having a 50% higher risk than the non-exposed [(relative risk $- 1$)*100]. This later interpretation responds to the relative increase of risk.

$$\text{Relative risk} = \frac{\text{Cumulative incidence among exposed}}{\text{Cumulative incidence among non-exposed}}$$

$$\text{Odds ratio} = \frac{\text{Odds of exposure among cases}}{\text{Odds of exposure among controls}} \text{ (Case-control studies)}$$

$$\text{Odds ratio} = \frac{\text{Odds of developing the disease among exposed participants}}{\text{Odds of developing the disease among non-exposed participants}} \text{ (Cohort studies)}$$

$$\text{Hazard ratio} = \frac{\text{Hazard function among exposed}}{\text{Hazard function among non-exposed}}$$

FIGURE 2.3   Measures of association.

Finally, when the risk among exposed is lower than the risk among non-exposed, the relative risk will have a value lower than 1. In this case, we will consider the exposure as a protective factor against the outcome that we are considering. For example, if the result of a study shows a relative risk of 0.75, this means that exposed participants have 0.75 times the risk of those not exposed. We could also say that the risk among exposed is 25% lower than among non-exposed [(1−relative risk)*100]. This latter interpretation corresponds to the preventive fraction.

2.3.2   Odds Ratio

The odds ratio is the measure of association that results from a case-control study, but sometimes it is also used in cohort studies (see later) [2].

In case-control studies, the odds ratio compares the odds of exposure among those who do have the disease (cases) to the odds of exposure among those who do not have the disease (controls) (Fig. 2.3).

If the exposure is equally frequent among cases than among controls, the odds ratio will have a value of 1. We will conclude that there is no association between the exposure and the disease.

If the exposure is more frequent among cases than among controls, then the odds ratio will have a value greater than 1 and this will mean that the exposure is a risk factor for the disease being considered.

Finally, if the exposure is less frequent among cases than among controls, the odds ratio will be lower than 1 and the exposure will be considered as a protective factor against the disease.

In cohort studies, the odds ratio is calculated as the ratio of the odds of developing the outcome among exposed participants over the odds of developing the outcome among non-exposed participants. Similarly to the odds ratio in case-control studies, an odds ratio of 1 means lack of association, an odds ratio greater than 1 is the result of a risk factor, and an odds ratio lower than 1 implies that the exposure is a protective factor.

2.3.3 Hazard Ratio

Hazard ratios are measures of association widely used in prospective studies (see later). It is the result of comparing the hazard function among exposed to the hazard function among non-exposed.

As for the other measures of association, a hazard ratio of 1 means lack of association, a hazard ratio greater than 1 suggests an increased risk, and a hazard ratio below 1 suggests a smaller risk.

We have presented here how to calculate the measures of association. What we obtain from these calculations is what we call the point estimates. The point estimates are the exact result that is to be obtained in a concrete study. Considering that the concrete result of a study accurately captures the population parameter of the association between an exposure and the outcome that we are studying can be somewhat risky. Therefore, measures of association are usually presented together with their 95% confidence interval. The 95% confidence interval tries to represent the reliable range of values in which we expect the true population parameter to be included.

2.4 WHAT TYPE OF STUDIES DO WE USE IN EPIDEMIOLOGY?

Epidemiological studies can be classified according to different study characteristics:

1. Purpose of the study: Based on the purpose of the study, epidemiological studies can be classified as:
   a. Descriptive studies: the aim of the study is to characterize a certain population.
   b. Analytic studies: the aim of the study is to assess if there is an association between a certain exposure and a certain outcome.
2. Exposure assignment: depending on how the exposure is assigned, studies can be classified as:
   a. Observational studies: the researchers do not assign the exposure. The study participants are exposed or not according to their personal traits, the surrounding environment, or their personal choices.
   b. Experimental studies: the researchers assign the exposure. Usually this is done randomly so that being exposed or not depends just on chance.

3. Follow-up time: information on the exposure and the outcome can be collected at one time point or exposure and outcome can be collected at different time points. Accordingly, studies can be classified as:
   a. Cross-sectional studies: the information is collected simultaneously at a certain point of time. Information on what is present at that point in terms of exposure and outcome is gathered.
   b. Longitudinal studies: there is a follow-up, so that there is a time span between the period of time when the exposure is happening and the time when the outcome is occurring.
4. Study direction: according to the study direction, the studies can be classified as:
   a. Cohort studies: the starting point is the classification of participants according to their exposure. Participants are then observed and new cases of disease among exposed and among non-exposed are registered.
   b. Case-control studies: the starting point is given by a group of participants who do have the disease (cases) and a group of participants who do not have the disease (controls). Information on their previous exposure to the factor of interest is collected.
5. Beginning of the study: depending on when the outcome occurs, we can distinguish:
   a. Retrospective studies: the outcome has already happened when the study is started.
   b. Prospective studies: the outcome has not occurred at study inception.
6. Unit of measurement: depending on the level of ascertainment, studies can be classified as:
   a. Individual studies: information is collected at the individual level.
   b. Ecological: information is registered at the population level.

2.4.1   Ecological Studies

Ecological studies are based on the analysis of group characteristics. If someone is interested in studying the association between red meat consumption and colorectal cancer, one possibility would be to conduct an ecological study collecting information on per capita supply of red meat in different countries and incidence rates of colorectal cancer in those same countries and establishing an association between those two characteristics [2].

The main limitation of the ecological studies is what is known as the ecological fallacy. Since only information on the average red meat consumption in each country and the incidence rate in each country are available, we cannot tell if those people with a higher meat intake were in fact those who developed colorectal cancer or not. Theoretically, it is possible that those consuming less red meat were those who experienced a higher risk of colorectal cancer.

2.4.2   Cross-Sectional Studies

In cross-sectional studies, all the information is collected simultaneously. There is no follow-up.

The main limitation of this type of study consists on the impossibility to assess what happened first. So, if we think that a certain trait may lead to a certain condition, since both have been collected at the same time we cannot tell accurately which one occurred first.

2.4.3   Case-Control Studies

The starting point in case-control studies is finding a group of participants with the disease that we are interested in studying (cases) and a group or participants with similar characteristics as the cases but who do not have the disease (controls) (Fig. 2.4).

In a second step, information on the previous exposure to the factor of interest is collected among cases and among controls.

Finally, the odds of exposure among cases is compared to the odds of exposure among controls.

If cases have already developed the disease when the study starts, then the study is called a case-control study with prevalent cases or retrospective case-control study.

If we include those cases that are just diagnosed while the study is going on, then the study is called a case-control study with incident cases or prospective case-control study.

2.4.4   Cohort Studies

In cohort studies, the recruited participants are first classified according to their exposure into exposed and non-exposed. Then, the incidence of the outcome of interest is ascertained in both groups. Finally, the incidence among exposed is compared to the incidence among non-exposed (Fig. 2.5).

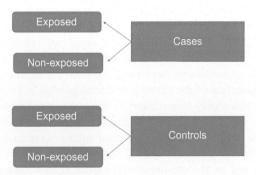

FIGURE 2.4   Design of case-control studies.

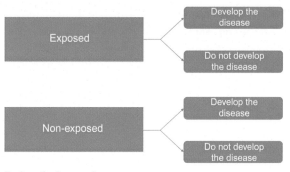

FIGURE 2.5 Design of cohort studies.

If the disease has happed before the beginning of the study, then the study is called a retrospective cohort study. If all participants are free of the disease when the study begins, then the study is called a prospective cohort study.

It is important that all participants are disease free by the time of being exposed or not.

2.4.5 Randomized Controlled Trials

The main difference between prospective cohort studies and randomized trials is that in the latter ones the researchers assign the exposure. The assignment is usually done at random for each participant who has been recruited (Fig. 2.6).

One group receives the intervention that is being tested and the other group receives a placebo or the usual care for the disease that is being studied.

Since the exposure is assigned at random, this design has the advantage that both groups—exposed and non-exposed—are equally balanced in their baseline characteristics, provided that the sample size is large enough. Thus, any differences that are subsequently observed between the two groups can be attributed to the exposure.

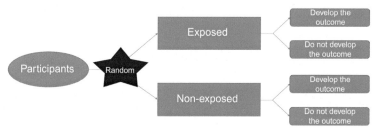

FIGURE 2.6 Design of randomized controlled trials.

2.4.6 Meta-Analysis

Available evidence on epidemiological studies is continuously growing. Sometimes it happens that more than one study has been published on a certain topic. When this is the case, it is possible to combine the results of the different studies and obtain one combined estimate that tries to summarize the available evidence on that topic. The technique that allows to do this is known as meta-analysis.

REFERENCES

[1] Rothman KJ. Epidemiology. An introduction. 2nd ed. New York: Oxford University Press; 2012.
[2] Gordis L. Epidemiology. 5th ed. Philadelphia, PA: Elsevier Saunders; 2014.
[3] Rosner B. Fundamentals of biostatistics. 6th ed. Belmont, CA: Thomson Higher Education; 2006.

Chapter 3

Not All Fats Are Unhealthy

Ligia J. Dominguez and Mario Barbagallo

3.1 INTRODUCTION

Fats, also called lipids (from the Greek = lipos meaning fat), are a heterogeneous group of substances that have in common a low degree of solubility in water and a low density. They are soluble in organic solvents such as benzene, ether, or chloroform. Dietary fats obtained by food consumption were recognized for the first time as an important nutrient in 1827 by William Prout. He proposed the classification of substances in food into sugars and starches, oily bodies, and albumen, which would later become known as carbohydrates, fats, and proteins [1]. It is essential to eat some fat, but overeating fat (as well as overeating other nutrients or foods containing them) may be harmful.

The consumption of daily dietary fat supplies the body with energy needed to function properly. During exercise, the body uses calories delivered from carbohydrates that have been consumed or stored in the body as glycogen. However, after twenty minutes, energy needed to continue exercising comes from calories derived from fat.

Fat is needed to build cell membranes (the exterior envelop of the cells), and to form the covers surrounding nerves making possible neural transmission. These compounds are necessary to maintain healthy hair and skin. Fats also consent the absorption of vitamins A, D, E, and K, which are called fat-soluble vitamins. They are also necessary for the absorption of other nutrients and phytochemicals from fruits and vegetables [2]. Fat also fills fat cells, called adipocytes, and represent a reservoir of energy needed for use in times of fasting or nonavailability of food. The fat deposited in adipocytes insulates the body to help keep body temperature warm. Fats are also involved in food palatability and they prevent body proteins (muscle) use as an energy source.

Fats obtained from food supply the body with essential fatty acids called linoleic acid and linolenic acid. They are called "essential" because the body cannot produce them by itself and does not function without them. The body

The Prevention of Cardiovascular Disease Through the Mediterranean Diet.
DOI: http://dx.doi.org/10.1016/B978-0-12-811259-5.00003-2

needs essential fatty acids for brain development, controlling inflammation, and blood clotting [3].

Fat supplies nine calories per gram, which corresponds to more than two-fold the number of calories provided by both carbohydrates and proteins, which have four calories per gram. Therefore, they have been called "fatten-ing." Nevertheless, we will discuss later how numerous studies do not entirely confirm that this is the case.

In recent decades, dietary recommendations have insistently claimed the reduction of fat intake. Initially, the recommendation to reduce fat intake was justified in seeking to reduce saturated fats prompting to lower choles-terol and thus theoretically decreasing the development of cardio- and cere-brovascular disease. But the campaign against saturated fats rapidly became a battle against all fats. It seemed reasonable to reduce fat also to fight obe-sity by its higher caloric content, which was undeniably not achieved because obesity continues to increase disproportionately and worryingly. For years, the differences between diverse types of fats were overlooked, only giving the general advice of decreasing total fat intake. This was an unwise decision also because the message to the lay public did not consider the key recommendation on how to replace the foods that should be avoided. Healthy food choices as alternatives should have been clearly suggested, which was absolutely neglected. The policy focusing on total fat reduction did not consider the harms of highly processed carbohydrates, such as refined grains, added sugar, and potato products, which consumption has been inversely related to that of total dietary fat. As a matter of fact, diets including an excess of refined carbohydrates are generally associated with alterations of blood lipids (i.e., high levels of triglycerides and low concen-trations of HDL or "good cholesterol") and other metabolic abnormalities associated with and elevated risk of heart attacks and strokes, including high blood glucose, insulin, and uric acid, increasing the risk for the development of diabetes and of fatty liver [4]. It is as well extremely worrying that school-lunch programs remain loaded with refined carbohydrates in an effort to reduce calories from total fat, instead of replacing unhealthy fats (i.e., industrial-made trans fat) with healthier fats, and with other healthy food choices [5]. Switching to a low-fat diet probably did not contribute to make people healthier, also because healthy fats with important heart and meta-bolic benefits, such as those contained in nuts, vegetable oils, and fish, were reduced as well as harmful ones. Therefore, dietary advice should put the emphasis on optimizing types of dietary fat and not on reducing total fat. Likewise, the advice to reduce refined carbohydrate intake should be linked to the advice not to replace the reduced calories with calories from unhealthy fat. Conversely, refined carbohydrates and sugar should be replaced by healthy carbohydrates such as whole grains, legumes, fruits, and vegetables, and also by healthy fats. The consumption of "low-fat," "nonfat," or "choles-terol-free" products, which generally contain high amounts of refined grains and added sugars, should be discouraged.

Therefore, for long-term health effects, some fats are better than others. Good fats include monosaturated and polyunsaturated fats. Bad ones include industrial-made trans fats, while saturated fats are in the middle. In this chapter, we will explain the differences among fats and the recommended dietary choices containing different types of fat.

3.2 WHAT IS FAT?

Dietary fat is called fat of animal or plant origin used as food. Occasionally, the term fat is used in reference to a solid lipid at room temperature (mostly in animal foods), compared to oils that are liquid at this temperature (derived from plants). However, "fat" is a generic term that is often used as a synonym for any form of lipid. In fact, all fats have a similar chemical structure: a chain of carbon atoms bonded to hydrogen atoms. What makes one fat different from another is the length and shape of the carbon chain and the number of hydrogen atoms connected to the carbon atoms. Apparently, slight differences in structure translate into key differences in form and function. In solid fats, saturated fatty acids predominate, while oils are predominantly unsaturated fatty acids.

The most common type of fat is one in which three fatty acids are bound to the glycerol molecule, receiving the name of "triglycerides," which have the shape of a small comb with only three teeth consisting of fatty acids or a capital letter E (Fig. 3.1).

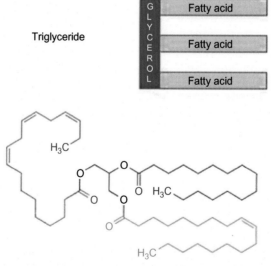

FIGURE 3.1 On top, schematic structure of triglycerides. Below, example of a triglyceride with three different fatty acids. One fatty acidis saturated (no double bonds) (*blue*), another contains one double bond within the carbon chain (*green*). The third fatty acid (a polyunsaturated fatty acid, *red*) contains three double bonds within the carbon chain.

Fatty acid

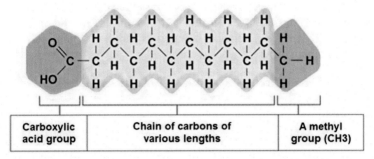

FIGURE 3.2 Schematic structure of a fatty acid. A chain of carbon and hydrogen atoms with a carboxyl group at the alpha end and a methyl group at the omega end.

A fatty acid consists of a chain of carbon (4−24) and hydrogen atoms with a carboxyl group at the alpha end and a methyl group at the omega end (Fig. 3.2).

Other classes of lipids are sterols (like cholesterol) and phospholipids (forming the cellular membranes).

Fats are formed from carbon, hydrogen, and oxygen, equal to carbohydrates, but the ratio between hydrogen and oxygen is much higher in fats. This is what makes them more energetic than carbohydrates in absolute terms. All types of fat provide the same number of calories (9 kcal/g) regardless of where they come from.

There are more than 500 types of fats, classified according to their molecular structure in simple, compound, and derivatives:

- **Simple lipids**: are the most abundant in the human body (approximately 95%) and in the diet (this form accounts for about 98% of the lipids present in food), as triglycerides. They represent the main form of deposit and use.
- **Lipid compounds**: triglycerides are combined with other chemical substance such as phosphorus, nitrogen, and sulfur. These compounds account for about 10% of body fat, including phospholipids, glycolipids, and lipoproteins.
- **Lipid derivatives**: derived from the processing of simple or compound lipids. The most important is cholesterol, but also include vitamin D, steroid hormones, palmitic, oleic, and linoleic acids.

Phospholipids are lipids containing phosphoric acid, present in the body and in some foods. They are part of cell membranes and various tissues, providing stability. Phospholipids are not particularly abundant in the diet, where they are present in animal viscera (e.g., liver, brain, and heart), and soybean and egg yolk. They are used in significant quantities as emulsifying additives (lecithin, E-322, allows mixing fat and water) to produce margarines, cheeses and other foods.

Cholesterol is a structural component of cell membranes in the body. Furthermore, other molecules of great functional importance are manufactured from it, such as vitamin D, steroid hormones, and biliary acids. That is, there is cholesterol that our body produces naturally and another that we get from food.

Cholesterol is transported in blood bound to proteins and other fats, forming the so-called lipoproteins. The best known are HDL-C (meaning High Density Lipoprotein cholesterol) or "good cholesterol" and LDL-C (meaning Low Density Lipoprotein cholesterol) or "bad cholesterol." HDL is considered good because it leads cholesterol from peripheral cells to the liver, preventing it from building up in the walls of blood vessels. The dietary cholesterol is only found in animal foods, among which viscera, meats and sausages, cream and butter, pastries and cakes containing animal fat.

Fatty acids, which are the building blocks of fat are classified as saturated, monounsaturated or polyunsaturated depending on their chemical structure (Fig. 3.3).

- **Saturated fatty acids**: are free of double bonds and therefore have the maximum number of hydrogen atoms. They are found not only in animal products (sausage, butter, meat, cheese, and cream) but also in plant foods (coconut and palm oil).
- **Unsaturated fatty acids**: contain one (mono) or more (poly) double bonds between the carbon atoms and the hydrogen.

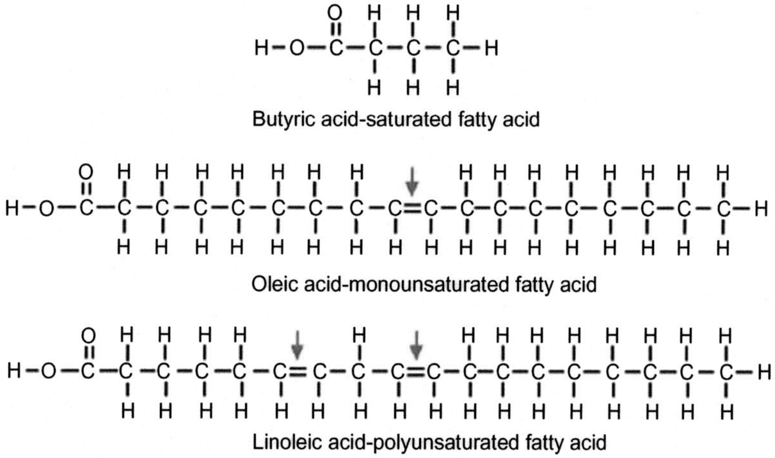

Butyric acid-saturated fatty acid

Oleic acid-monounsaturated fatty acid

Linoleic acid-polyunsaturated fatty acid

FIGURE 3.3 Chemical structure of a saturated (butyric), a monounsaturated (oleic), and a polyunsaturated (linoleic) fatty acid.

- **Monounsaturated fatty acids**: contain a single double bond between the carbon atoms that compose them. They are mostly found in olive oil and nuts.
- **Polyunsaturated fatty acids**: contain more than two bonds between the carbon atoms. They are contained in fish, nuts, vegetable oils (i.e., sunflower and corn) and in some plant extracts. These fats include essential and semi essential fatty acids:
  - Essential fatty acids, alpha linolenic, and linoleic acids: they cannot be synthesized by the human body; are the precursors of prostaglandins, thromboxanes, and leukotrienes, substances that are involved in the function of immune system, inflammatory response, and affect the cardiovascular system (Fig. 3.4).
  - Semi essential fatty acids, including EPA and DHA (Fig. 3.5).

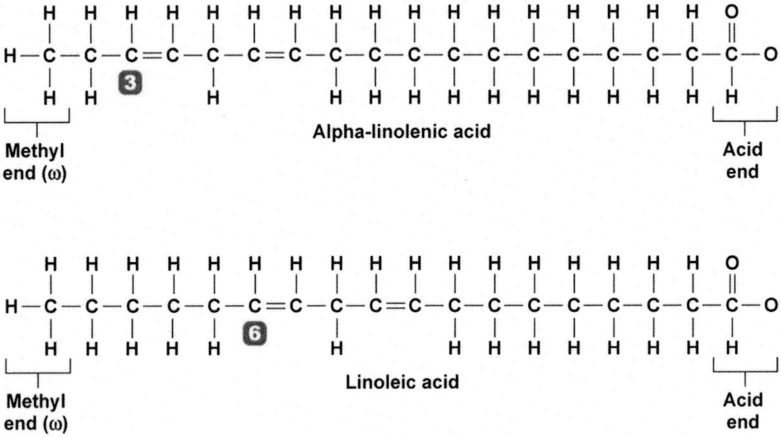

FIGURE 3.4   Chemical structure of essential fatty acids: alpha-linolenic and linoleic acids.

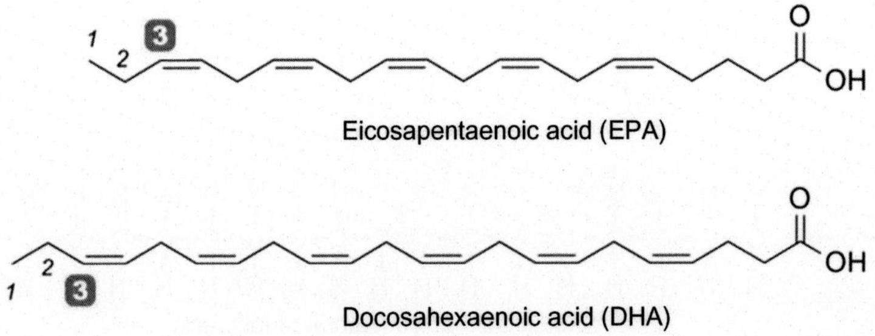

FIGURE 3.5   Chemical structure of omega-3 fatty acids: eicosapentaenoic and docosahexaenoic.

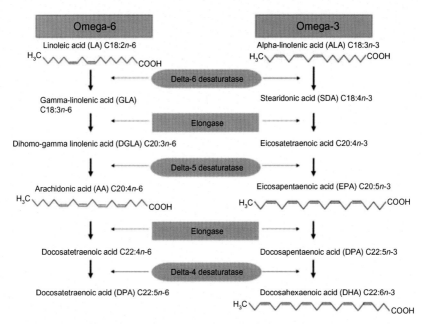

FIGURE 3.6 Metabolic pathways of linoleic acid and alpha-linolenic acid, which compete for the same enzymes biosynthetic pathways.

They are called semi essential because they can derive from the biosynthetic pathway of linoleic acid and alpha-linolenic acid, through distinct metabolic pathways, which compete for the same enzymes (Fig. 3.6).

Particular interest has been shown toward this category of enzymes, called desaturase, because they can be reduced in many conditions, that is, diets rich in trans fat, stress, drastic diets, malabsorption, diabetes, ionized radiation, cancer, aging, and deficiency or malabsorption of fat-soluble vitamins. Therefore, the introduction of these fatty acids (i.e., EPA and DHA from fish or fish oil) is able to bypass the problem linked to the possible deficiency of Δ-6-desaturase. This is why fish and fish oil are considered a better source compared to oil and flaxseed, which are rich in their precursor alpha-linolenic acid. EPA and DHA are very important precursors of substances essential for the body's health.

- **Hydrogenated fatty acids**: as mentioned, normally the acids of vegetable fats are liquid at room temperature. They can be rendered solid by the hydrogenation process, which alters the chemical structure making them particularly harmful for our health. They are the so-called trans-fatty acids or hydrogenated fatty acids (Fig. 3.7).

Triglycerides represent the form of storage of fatty acids in the body (think of the fat accumulated in the belly and everywhere else). During the

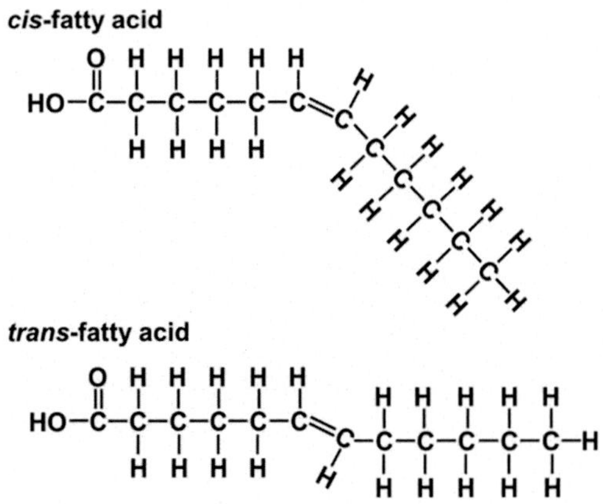

FIGURE 3.7 Chemical structure of a *cis*-fatty acid and a *trans*-fatty acid conformation.

energy processes, the body provides to split the bond between glycerol and fatty acids channeling them into two completely different metabolic pathways. While glycerol is used to produce glucose, free fatty acids are transported in the bloodstream in association with albumin, a plasma protein that transports them to the muscles, where they are energy substrate for oxidative processes.

3.3 FAT ABSORPTION

When fat from foods reaches the small intestine, an enzyme (protein that breaks down chemicals) called pancreatic lipase digests the fatty acids, giving rise to so many fine aggregates called micelles.

Within these small "carriers," needed to transfer lipophilic molecules in the cells responsible for their absorption, are the products of lipid digestion, such as cholesterol, vitamins, bile salts, monoglycerides, and fatty acids resulting from digestion of triglycerides and of phospholipids. The micelles are soluble in aqueous environment due to the compact size and solubilizing action of bile salts. Arrived in the outer surface of intestinal villi or brush border (finger-like projections extended into the lumen), near the microvilli (projections from each cell), the micelles release their contents. The individual components, by virtue of their lipophilicity, manage to cross the plasma membrane of the brush and to penetrate the intestinal cells (enterocytes).

In the cytoplasm of the enterocyte fatty acids combine to reform triglycerides (exactly the reverse of what happened in the stomach and especially in the initial portions of the small intestine). At this point, it is produced as a lipoprotein, called chylomicron (Fig. 3.8), which consists of a lipid core (formed by triglycerides, phospholipids, cholesterol, and vitamins), surrounded by protein molecules. This sort of mantle, due to the hydrophilic

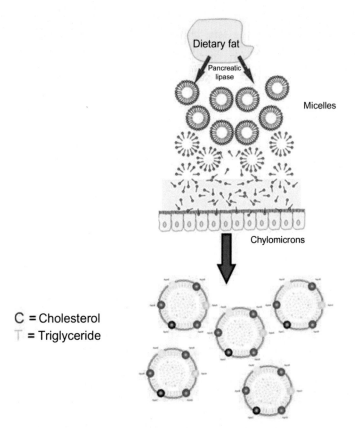

C = Cholesterol
T = Triglyceride

FIGURE 3.8 Scheme of intestinal fat absorption through micelles and chylomicrons (one type of lipoprotein). Apolipoproteins (A, B, C, E) are proteins that bind lipids (i.e., triglycerides and cholesterol) to form lipoproteins (fat carriers, such as in other lipoproteins, good [HDL] and bad [LDL] cholesterol).

nature conferred by the protein, increases the degree of solubility of chylomicron in the aqueous medium. In this manner, chylomicrons exit enterocytes, pass into the fluid surrounding the cells (interstitial fluid) and from here to inside the lymphatic vessels. Bile acids and salts are instead absorbed in the terminal small intestine (ileum), transmitted into the bloodstream and transported to the liver, where they are recycled and resecreted in bile. After chylomicrons are conveyed in the lymphatic system, they are transferred into the bloodstream in the subclavian veins (below collarbone). These large agglomerations, by virtue of their large size, would encounter many difficulties to cross the blood capillaries inside the villus. In summary, the chylomicrons pass into the blood only after having been absorbed in the intestine and transported by the lymph to the blood. All other intestinal absorption products (carbohydrates, amino acids, water, minerals, and water-soluble vitamins) enter directly into blood capillaries by simple diffusion, facilitated or active transport. Among fats, only short- and medium-chain fatty acids, which represent only a small part of lipids in food, go directly to the blood capillaries.

3.4   GOOD FATS

Fats with health benefits are mainly of plant origin, such as vegetable oils, nuts, and seeds. Also fatty fish contains healthy fats. The two main categories of fats with beneficial health effects are monounsaturated and polyunsaturated fats. As mentioned earlier, the difference with saturated fats is that unsaturated fats have less hydrogen atoms linked to the chain of carbon atoms.

3.4.1   Monounsaturated Fats

The first suggestion that monounsaturated fatty acids are associated with beneficial effects comes from the Seven Countries Study in the 1960s. In this multicenter study, it was observed that people from Greece after World War II and in other parts of the Mediterranean basin, had a low frequency of heart disease despite high fat intake in the diet. The major source of fat in their diet came from olive oil, which is rich in monounsaturated fatty acids, as opposed to saturated animal fat, commonly consumed in nations with higher rates of heart disease [6]. These findings started an ever-growing interest in Western countries on olive oil and on knowing how to adopt the healthy food traditions of people from the Mediterranean countries (e.g., Italy, Spain, Greece, France, and North Africa). The Mediterranean diet is regarded as one of the most healthful choices today, based on numerous studies confirming its multiple health benefits, in particular when supplemented with olive oil and nuts [7]. Foods typical of the Mediterranean diet include fresh vegetables and fruit, nuts, legumes, whole grains, herbs and seed, eggs, red wine with meals, and infrequent consumption of meats. A large clinical trial from Spain found that a Mediterranean diet with more nuts and extra-virgin olive oil, optimal sources of good fats, reduced heart attacks, and strokes [8] as well as diabetes [9], when compared with a lower fat diet with more starches. Olive oil has been used for centuries in the Mediterranean basin (Fig. 3.9).

Monounsaturated fats have a single carbon-to-carbon double bond (Fig. 3.3) and the most representative of these fatty acids is oleic acid, contained in olive oil. The result is that it has two fewer hydrogen atoms than a saturated fat and a bend at the double bond. This structure keeps monounsaturated fats liquid at room temperature. These good fats protect the cardiovascular system because they reduce levels of total blood cholesterol at the expense of so-called bad cholesterol (LDL) and increases good cholesterol

FIGURE 3.9 Centenarian olive tree from Modica (Sicily).

(HDL) [10]. Good sources of monounsaturated fats are olive oil, avocado, olives, peanut oil as well as most nuts.

It has been suggested that modifications of dietary components may amend the structure and function of cell membranes and, hence, counteract the damage induced by free radicals. Accumulation of free radicals, which are waste products of cellular metabolism, is called "oxidative stress" and it has been linked to aging and to the development of the most common chronic diseases, such as, heart disease, diabetes, dementia, and other neurodegenerative diseases. Olive oil is a natural product, particularly rich in monounsaturated fatty acids, mainly oleic acid. Monounsaturated fatty acids have an excellent fluidity and are less likely to undergo oxidation. Other elements contained in extra-virgin olive oil are antioxidant molecules, such as alpha tocopherol, phenolic compounds, and coenzyme Q, all of which contribute to counteract the toxic effects of free radicals protecting the cells from oxidative damage [11] (Table 3.1).

The recognized protective actions of olive oil on heart disease had been in the past attributed mostly to its high content of monounsaturated fatty acid. However, oleic acid is also present in some products of animal origin. Thus, in addition to the beneficial effects of oleic acid the powerful antioxidants (polyphenols, antioxidant vitamins, and coenzyme Q) contained in extra-virgin olive oil may help to explain its multiple benefits.

TABLE 3.1 Chemical Composition of Extra-Virgin Olive Oil

Glycerides	Minor Compounds
Triacylglycerols	*Hydrocarbons:*
Diacylglycerols	• Squalene
Monoacylglycerols	
Fatty acids:	*Alcohols:*
• Miristic	• Triterpenic
• Palmitic	• Alifatic
• Palmitoleic	
• Margaric	*Sterols:*
• Eptadecanoic	• β-Sitosterol
• Estearic	• Δ-5-avenansterol
• Oleic	• Campesterol
• Linoleic	• Δ-7-stigmasterol
• Linolenic	
• Arachic	Phenols
• Eicosanoic	Tocopherols
• Behenic	Phospholipids
• Lignocenic	Volatile compounds
	Pigments:
	• Chlorophylls A and B
	• Pheophytins A and B
	• β-Carotene
	• Lutein

Polyphenols are a large group of molecules present in all plants, essential for plant growth, reproduction, nutrition, and protection against some pathogens. They also give diverse colors to plant products. There are multiple types of polyphenols but the most frequent are the flavonoids, acid phenols, tannins, lignans, coumarins, stilbenes, and secoiridoids. On average, the phenolic compounds reach concentrations of 40−900 mg/kg in extra-virgin olive oil, while refined olive oil has on average 0.5 mg/kg (Table 3.2).

When considering all monounsaturated fats, the evidence of protection against heart disease is mixed [12]. As mentioned earlier, monounsaturated fatty acids may be also present in some foods of animal origin and this may contribute to the uneven results. However, when saturated fat is replaced with monounsaturated fat, it has been shown that they lower glucose in persons predisposed to diabetes [13]. Different food sources of monounsaturated fats can have different health effects. For example, some studies suggest that only the consumption of olive oil, but not that of all monounsaturated fatty acids, considering animal and plant sources, show benefit [14]. This confirms that other compounds, such as the antioxidants mentioned earlier, may help to explain the benefits of extra-virgin olive oil as well as of nuts. Thus, the Institute of Medicine recommends using monounsaturated fat along with polyunsaturated fats to replace saturated and trans fats.

TABLE 3.2 Phenolic Compounds of Extra-Virgin Olive Oil (20—900 mg/kg)

Phenolic Acids and Derivatives	Secoiridoids
<ul><li>Vanillic acid</li><li>Syringic acid</li><li>p-Coumaric acid</li><li>o-Coumaric acid</li><li>Gallic acid</li><li>Caffeic acid</li><li>Protocatechuic acid</li><li>p-Hidroxybenzoic acid</li><li>Ferulic acid</li><li>Cinnamic acid</li><li>4-(acetoxyethil)-1,2-Dihydroxybenzene</li><li>Benzoic acid</li></ul>**Phenolic alcohols**<ul><li>Hydroxytyrosol</li><li>Tyrosol</li></ul>	<ul><li>Dialdehydic form of decarboxymethyl elenolic acid linked to 3,4-DHPEA (3,4-DHPEA-EDA)</li><li>Dialdehydic form of decarboxymethyl elenolic acid linked to p-HPEA (p-HPEA-EDA)</li><li>Oleuropein aglycon (3,4-DHPEA-EA)</li><li>Ligstroside aglycon</li><li>Oleuropein</li><li>Dialdehydic form of oleuropein aglycon</li><li>Dialdehydic form of ligstroside aglycon</li></ul>**Lignans**<ul><li>(+)-1-Acetoxypinoresinol</li><li>(+)-Pinoresinol</li></ul>**Flavones**<ul><li>Apigenin</li><li>Luteolin</li></ul>**Hydroxy-isocromans**

3.4.2 Polyunsaturated Fats

Polyunsaturated fatty acids have two or more double bonds in its carbon chain, are necessary to build cell membranes and the covering of nerves as well as for proper blood clotting, muscle movement, and inflammation. There are two main types of polyunsaturated fats: omega-3 fatty acids and omega-6 fatty acids. The numbers denote the distance between the beginning of the carbon chain and the first double bond (Figs. 3.3 and 3.4). Both types provide health benefits. The most common are omega-6 linoleic acid and omega-3 alpha-linolenic acid, derived principally from plants and vegetable oils (e.g., flaxseeds, nuts, safflower oil, and unhydrogenated soybean oil).

Fatty fish (i.e., salmon, mackerel, and sardines), is a major source of long-chain omega-3 polyunsaturated fats, in particular EPA and DHA (Fig. 3.5). Linoleic (omega-6) and alpha-linolenic (omega-3) fatty acids are essential, which means that they are necessary for normal body functions but the body can't make them. Thus, they must be consumed with food. This is in contrast with saturated and monounsaturated fat that can be synthesized in the liver. Humans synthesize also little amounts of EPA and DHA, whereby diet remains the major source of these essential fatty acids.

The positive effects for preventing or even treating heart disease and stroke with fish and omega-3 consumption have been extensively studied.

Moderate consumption of fish ($\sim 2-3$ servings/week) and of long-chain omega-3 (~ 250 mg/d) reduces the risk of fatal heart attack compared to little or no consumption [15,16].

The reduction in stroke and nonfatal heart attacks with fish consumption or supplementation with fish oil is present but less strong. Interestingly, earlier studies gave positive results, while newer studies are negative [17]. These discrepancies could be due to the more aggressive therapies to decrease blood lipids and hypertension in recent years or by general higher consumption of fish, which conceals the effect of added fish.

Fish and omega-3 consumption have been shown to improve major risk factors, such as hypertension, triglycerides, inflammatory markers, arterial stiffness, and cardiac function [15]. The effects of fish consumption on heart failure, atrial fibrillation, and cognitive decline remain unclear.

There is some concern that possible fish contaminants (e.g., polychlorinated biphenyls, dioxins, and mercury), can reduce the beneficial effects. For example, polychlorinated biphenyls have been associated with increased heart attacks [18] and hypertension [19]. The use of a variety of fish avoiding the frequent consumption of large fish is encouraged to prevent the accumulation of contaminants. For pregnant women, the FDA recommends $2-3$ servings/week of a variety of fish lower in mercury, while avoiding specific species (Gulf of Mexico tilefish, shark, swordfish, king mackerel; albacore tuna up to 6 oz/week) [20].

Replacing saturated fats or highly refined carbohydrates with polyunsaturated fats reduces the risk of heart attack or cardiac death [21]. Dietary linoleic fatty acid (omega-6) reduces LDL (bad cholesterol), triglycerides, and is also associated with significantly lower risk of heart disease [22]. Therefore, polyunsaturated fat-rich vegetable oils possibly ought to replace refined starches and added sugars.

3.5 BAD FATS

3.5.1 Trans Fats

The worst type of dietary fats and widely accepted as being unhealthy for good reason is the so-called trans fat, which has one or more double bonds in a trans position, rather than the natural cis position (Fig. 3.7). In nature, *cis*-fatty acids are the most common compared to trans fat, which are formed mainly due to certain artificial treatments. Naturally occurring trans fats such as those found in grass-fed meat, are quite different and exist at a very low concentration compared to industrially produced trans fats.

Trans fat is the final product of a chemical process called hydrogenation which is used to transform healthy monounsaturated and polyunsaturated vegetable oils into solid and prevent them from becoming rancid. Trans fats were created by the food industry because they are more stable fats and have a longer shelf life. Hydrogenation also makes them more easily spreadable and modifies the flavor of baked goods. The industrial process of hydrogenation consists of heating oils in the presence of hydrogen and a heavy-metal catalyst, such as palladium, nickel, cobalt, copper, or iron, changing the chemical structure of oils and causing the oil to become a solid. On food label ingredient lists, this manufactured substance is typically listed as "partially hydrogenated oil." The amount of trans fat in partially hydrogenated vegetable oils is substantial accounting for about 30%−60% of trans fat.

At the end of the 19th century Paul Sabatier, a Nobel laurate, developed the chemical process of hydrogenation of vapors [23], which was developed for liquid oils by the German chemist Wilhelm Normann at the beginning of the 20th century, calling it "fat hardening". The procedure was used in the next years to start the industrial production of "solid oils" at large scale in England. The extend of the return is exemplified by the production of near 3000 tons in the first year of production at Joseph Crosfield & Sons. In 1909, Procter & Gamble acquired the US rights to Normann's patent and 2 years later, it started the first hydrogenated shortening, Crisco, which was widely publicized by giving free cookbooks with various recipes made with this shortening. Hydrogenation allowed the use of inexpensive oils, which were not well received by consumers such as whale oil or fish oil, a practice kept secret. Also in the early 20th century, the United States started soybeans imports as a source of proteins. Soybean oil was a by-product that rapidly turns to be a source for the industry of shortenings, replacing butterfat, which was not sufficiently available for the growing numbers of eager consumers. Soon after, the industry took advantage on the properties of hydrogenated fats, which allowed the possibility, unlike butter, to be taken out of the refrigerator and immediately spread on bread, possibly triggering the frequent and excessive use. This product also provided gainful baking properties, hence, soybean oil-made margarine started replacing butter and lard in the production of breads, pies, cookies and cakes in the 1920s. The production of hydrogenated fats increased steadily until the 1960s replacing animal fats in all Western countries, claiming the convenience of lower costs and even sold as healthier than saturated fats of butter [24].

The first suggestions in the scientific literature on the contribution of trans fat to the large increase in heart disease started in the mid-1950s. However, this concern remained ignored. Fats of animal origin began to be a concern by the 1980s, which resulted in the switching to trans fat in most fast food and other food companies. Several studies at the beginning of the 1990s confirmed the negative health impact of trans fats, estimating that trans fat had caused 20,000 deaths annually in the United States from heart disease [25]. Mandatory food labeling for trans fat was thus introduced in several countries constraining food companies to change their manufacturing practices to meet the FDA definition of "0 g trans fat per serving," which corresponded to less than 1 g per tablespoon, or less than 0.5 g per serving size. However, the exclusion of trans fats in food manufacturing has taken longer than expected. Not only because of resistance of food industry to change their convenient formulations, but also because often taste and food gratification take precedence over perceived risk to health among laypeople.

Thus, during the 20th century, trans fats went from being present only in margarines and vegetable shortening to the widespread use in industrial food production because of financial convenience, wholly neglecting the dire health effects of hydrogenated oils. This resulted in their widespread presence in everything from margarine, commercial cookies, pastries, frozen foods, fried foods, packaged snacks to French fries in fast food. Trans fats have been found to be incorporated into both fetal and adult tissues [26]. Their negative health consequences are well understood and documented.

Higher trans fat intake from partially hydrogenated oils and the myriad of products made using them are consistently associated with an elevated risk of heart disease and of sudden death in an intake-dependent way [27]. Trans fat has also multiple adverse effects, such as raising deleterious LDL, triglycerides, and lowering beneficial HDL. These effects are true whether trans fat replaces saturated, monounsaturated of polyunsaturated fat [28]. Trans fat also appears to promote inflammation, vascular alterations, and arrhythmia. They contribute to insulin resistance and visceral obesity, which increases the risk of developing type 2 diabetes [29]. Even small amounts of trans fat can be harmful: for every 2% calories from trans fat consumed daily, the risk of heart disease rises by 23%. Trans fats have no known health benefits and there is no safe level of consumption. Therefore, their consumption should be avoided completely.

Today, the food industry is proposing to use alternative technologies to hydrogenation, with the goal of obtaining vegetable fat devoid of hazardous *trans*-fatty acids, but with the same organoleptic characteristics. However, these are artificially manipulated products, not natural, and maybe made from low-quality oils or already rancid. In addition, they have a high content of saturated fatty acids, because semi-solid at room temperature.

It was not until 2006 that consumers in the United States could directly be aware of the presence of trans fat in food, although only generally labeled as "partially hydrogenated" ingredients. In 2003, an FDA regulation asked manufacturers to list trans fat on the Nutrition Facts panel of foods and some supplements. Trans fat levels of less than 0.5 g per serving can be listed as 0 g trans fat on food labels. However, 0.5 g per serving threshold could be too high for persons eating many servings of a trans fat-containing product or multiple products during the day, consuming significant amounts of trans fat. Furthermore, without specific prior knowledge about trans fat and its negative health effects, consumers, including those at risk for heart disease, may misinterpret nutrient information provided on the panel. There is no requirement to list trans fat on bulk distribution for schools, hospitals, jails, and cafeterias.

At present, the need of elimination of these industrial man-made fats is considered a public health priority. In the meantime, it is crucial to recognize them through attentive food label reading.

3.6 IN-BETWEEN FATS

3.6.1 Saturated Fats

As mentioned, the term saturated means that this type of fat does not have double bonds in their structure because their carbon chain is saturated with hydrogen atoms. Saturated fats, which are solid at room temperature, are common in the diet because they are present in a variety of dissimilar foods, so it is likely that the effects of these fats, and foods containing them, are highly heterogeneous. The extension of the carbon chain in saturated fats ranges from 4 to 24 carbons and they have remarkably diverse biological features. For example, palmitic acid with 16 carbons has shown adverse effects, while medium (6−12 carbons), odd-chain (15−17 carbons), and very long-chain (20−24 carbons) fatty acids have shown beneficial metabolic effects [30,31]. Thus, it is probably unwise to put all saturated fats in one single category.

Likewise, the effects of saturated fats on cholesterol and other lipids in the blood are very complex and variable. It is known that a diet rich in saturated fats can raise "bad" cholesterol (LDL), which favors the formation of blockages in arteries in the heart and elsewhere in the body, but in some cases there is evidence that they may also increase "good" cholesterol (HDL) [32], which does the opposite. There are also studies showing that some saturated fats can contribute to lowering triglycerides, which could mean benefit. These mixed effects have raised animated discussions on the convenience or not of including or avoiding foods containing saturated fat in a healthy diet.

Dietary saturated fats are present in so many different types of foods such as cheese, chicken, butter, desserts with dairy products, sour cream, processed meats or sausages (i.e., ham, hamburgers, hot dogs, salami, corned beef, mortadella, and bacon), unprocessed red meat, viscera, lard, liver pâté, eggs, snacks, industrial bakery products, milk, yogurt, vegetable oils including coconut oil, and even in nuts. All these foods contain other elements besides saturated fats that can greatly modify their health effects. Perhaps this is the reason why there are studies with paradoxical and contradictory findings when taking into account only nutrients instead of foods with various nutritional components, or combinations of foods in dietary patterns. What's more, blood levels of some saturated fats with adverse effects (e.g., those with 14−16 carbons) come from their production in the liver in response to eating starches and sugar [33]. Also, levels of saturated fats in the blood correlate more with the consumption of starches and added sugars than with the consumption of meat and dairy [30].

This complexity clarifies why total consumption of saturated fats has little connection with measures of health. Indeed, many large and exhaustive studies found no evidence that eating saturated fat increased heart attacks and other cardiac events [34,35]. For example, highly processed meats, very rich in sodium, are more strongly associated with heart disease than the total saturated fat content of the diet [36]. Cheese, a primary source of saturated fat, has been linked in some studies to neutral or even beneficial effects on heart disease and diabetes [37]. These findings have challenged the accepted wisdom that saturated fat is inherently bad and will continue the debate about what foods are best to eat. Nevertheless, these studies should not be taken as "a green light" to eat more sausages, steak, butter, and other foods rich in saturated fat. Looking at individual fats and other nutrient groups in isolation could be misleading, because when people cut down on fats they tend to eat more bread, cereals, and other refined carbohydrates that can also be bad for cardiovascular health. The latest dietary guidelines put more emphasis on real food rather than giving an absolute upper limit or cutoff point for certain macronutrients [12].

Indeed, the beneficial effects described for some foods that contain saturated fat like nuts or some vegetable oils [8,38] cannot support benefit of all foods containing saturated fat. For example, there is no study showing that processed meat is protective for heart disease and instead there are studies showing that processed meat consumption increases the risk of heart attacks and diabetes when consumed in excess [39—41]. Yet, recent reports from the media have mistakenly mixed the complexity of positive or neutral effects of some foods containing saturated fat or some specific fatty acids (together with other protective components) with unproven recommendations to eat more butter and bacon as neutral or even beneficial for health.

Even if the aforementioned studies have shown no link between saturated fat and heart disease, there is evidence that replacing saturated fat with polyunsaturated fat or with whole grains may indeed reduce the risk of heart disease [21,22]. Furthermore, replacing saturated fat with polyunsaturated fats like vegetable oils or high-fiber carbohydrates is the best bet for reducing the risk of heart disease, but replacing saturated fat with highly processed carbohydrates could do the opposite [42].

3.7 ALL TYPES OF FATS ARE PART OF FOODS AND DIETARY PATTERNS

Unquestionably, fat consumption cannot be considered in isolation because people do not eat only fat but foods and food combinations or patterns of food [43]. The relationship between diet and chronic disease cannot be adequately predicted from the effects of individual nutrients, so focus should be put not so much on individual nutrients but rather on foods and dietary patterns with their joined and synergistic actions, also because people select foods, not nutrients, when deciding what to eat. In fact, focusing on single nutrients often leads to paradoxical dietary choices and industry formulations, and contributes to confusion about what constitutes a healthy diet. Few people understand or can correctly estimate their daily consumption of nutrients such as calories, cholesterol, fats, salt, fiber, or vitamins. Furthermore, specific foods and overall diet patterns, rather than single isolated nutrients, are most relevant for health.

Ours is a time of nutritional crisis, in which there is a lot of information on the components of diet, deconstructing it up to single nutrients, and various data on their positive or negative effects. But the crisis is manifested conspicuously in the form of the obesity epidemic that threatens to reverse the gains in life expectancy achieved in the last century. The obstinacy on limiting total fat intake to less than 35% has no clear evidence. Nutritional recommendations should be directed to limit the excess of food consumed beyond the need for body functioning rather than advice the reduction of the portion of calories from fat, because many researches in the field nowadays think that this recommendation does not have a clear scientific basis. Moreover, within the United States, declines in the percentage of energy from fat in the last decades have paralleled a massive increase in obesity, and similar trends are occurring worldwide in industrial nations. Thus, dietary fat not necessarily accounts for the high prevalence of excess body fat. Conversely, reduction in fats from the diet may further exacerbate this problem because it may be a relevant distraction from other helpful efforts to control obesity and improve health.

The persistent limit in total fat intake to 35% derived from a 2002 study that indicated the alleged association of the percentage of dietary fat with obesity [44]. However, this relationship has been refuted by several controlled, prospective, observational studies, and clinical trials that show a very small and no significant independent effect of dietary fat on body weight [12]. Notwithstanding, the diet of millions of people continue to be loaded with refined carbohydrates in an attempt to reduce fat, while the real goal should be to replace trans fats for healthy fats and for other healthy food choices. In addition, even if former RDAs (recommended daily allowances), now DRIs (dietary reference intakes), may have some general consideration, the recommendations about foods and dietary patterns should be based primarily on evidence on health effects and not on levels of specific nutrients [45].

The formulation of nutritional guidelines can sometimes have upsetting consequences. For example, it is curious to note that the original nutritional pyramid proposed in 1992 [46] in the Dietary Guidelines for Americans, which recommended replacing dietary fat with carbohydrates, without specifying their type or nutritional quality, may have contributed inadvertently to the epidemic of metabolic syndrome and associated chronic diseases with the understated message of increasing the consumption of refined carbohydrates. Regrettably, the focus on limiting total fat did not account for the health benefits of plant-derived fats or the harms of processed carbohydrates, which are indeed the most common replacement when dietary fat is reduced.

In recent years, high-fat diets have been at least as effective as low-fat diets in trials of short-term weight loss [12]. Weight maintenance in the long term is not associated with the fat content of diet and do not allow differentiating protective vs. harmful effects [47,48]. Low-fat diets have not shown benefits for major chronic diseases in large observational and controlled trials [27,49]. The large Women's Health initiative study did not show benefits for any major endpoint including heart disease, stroke, cancers, diabetes, or insulin resistance by lowering total dietary fat among nearly 50,000 US women followed for about ten years [28,34,50,51]. Conversely, controlled trials have shown that diets higher in healthful fats, exceeding the current 35% limit, reduce markers of heart disease risk and diabetes [52,53], and the Mediterranean diet, also exceeding the 35% dietary fat limit, actually decreased heart disease and diabetes [8,9]. These studies reaffirm the wisdom that various traditional diets in the world, rich in fats from vegetable oils, nuts, seeds, and seafood, together with fruits, vegetables, legumes, whole grains, yogurt, and with fewer red and processed meats, refined starches, and added sugar are the healthiest choices. Such diets are rich in fiber, antioxidants and they do not include man-made trans fat. The Mediterranean dietary pattern is a paradigm of such type of diet [7,12,54]. Based on all the aforementioned evidence, the new 2015 Dietary Guidelines state that recommendations should not focus only on lowering total fat [12]. Yet, the restriction on total fat still shapes numerous US nutritional programs and policies [55], and pushes industry production and

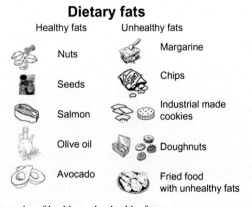

FIGURE 3.10 Examples of healthy and unhealthy fats.

marketing of low-fat snacks, desserts, biscuits, salad dressings, processed meats, and other products of poor nutritional value, replacing removed fats with refined starches and added sugars of the lowest quality, and falsely convincing people that avoiding fats and consuming these low-quality products is the best way to follow a healthy diet.

In summary, fats in the diet are necessary, so they should not be reduced at all costs. Nevertheless, excessive consumption of foods rich in fats is not advisable (as with any other food), and healthier fats should be preferred as an alternative of harmful ones (Fig. 3.10).

REFERENCES

[1] Rosenfeld L. William Prout: early 19th century physician-chemist. Clin Chem 2003;49:699–705.

[2] Brown MJ, Ferruzzi MG, Nguyen ML, Cooper DA, Eldridge AL, Schwartz SJ, et al. Carotenoid bioavailability is higher from salads ingested with full-fat than with fat-reduced salad dressings as measured with electrochemical detection. Am J Clin Nutr 2004;80:396–403.

[3] Das UN. Essential Fatty acids—a review. Curr Pharm Biotechnol 2006;7:467–82.

[4] DiNicolantonio JJ, Lucan SC, O'Keefe JH. The evidence for saturated fat and for sugar related to coronary heart disease. Prog Cardiovasc Dis 2016;58:464–72.

[5] Hopkins LC, Gunther C. A historical review of changes in nutrition standards of USDA Child Meal Programs relative to research findings on the Nutritional Adequacy of Program Meals and the Diet and Nutritional Health of Participants: Implications for Future Research and the Summer Food Service Program. Nutrients 2015;7:10145–67.

[6] Keys A. Mediterranean diet and public health: personal reflections. Am J Clin Nutr 1995;61:S1321–3.

[7] Martinez-Gonzalez MA, Salas-Salvado J, Estruch R, Corella D, Fito M, Ros E. Benefits of the Mediterranean Diet: insights from the PREDIMED study. Prog Cardiovasc Dis 2015;58:50–60.

[8] Estruch R, Ros E, Martinez-Gonzalez MA. Mediterranean diet for primary prevention of cardiovascular disease. N Engl J Med 2013;369:676–7.

[9] Salas-Salvado J, Bullo M, Estruch R, Ros E, Covas MI, Ibarrola-Jurado N, et al. Prevention of diabetes with Mediterranean diets: a subgroup analysis of a randomized trial. Ann Intern Med 2014;160:1–10.

[10] Lamarche B, Couture P. Dietary fatty acids, dietary patterns, and lipoprotein metabolism. Curr Opin Lipidol 2015;26:42–7.

[11] Servili M, Esposto S, Fabiani R, Urbani S, Taticchi A, Mariucci F, et al. Phenolic compounds in olive oil: antioxidant, health and organoleptic activities according to their chemical structure. Inflammopharmacology 2009;17:76–84.

[12] Millen BE, Abrams S, Adams-Campbell L, Anderson CA, Brenna JT, Campbell WW, et al. The 2015 Dietary Guidelines Advisory Committee Scientific Report: development and major conclusions. Adv Nutr 2016;7:438–44.

[13] Schwingshackl L, Strasser B, Hoffmann G. Effects of monounsaturated fatty acids on glycaemic control in patients with abnormal glucose metabolism: a systematic review and meta-analysis. Ann Nutr Metab 2011;58:290–6.

[14] Schwingshackl L, Hoffmann G. Monounsaturated fatty acids, olive oil and health status: a systematic review and meta-analysis of cohort studies. Lipids Health Dis 2014;13:154.

[15] Mozaffarian D, Wu JH. Omega-3 fatty acids and cardiovascular disease: effects on risk factors, molecular pathways, and clinical events. J Am Coll Cardiol 2011;58:2047−67.

[16] Rizos EC, Ntzani EE, Bika E, Kostapanos MS, Elisaf MS. Association between omega-3 fatty acid supplementation and risk of major cardiovascular disease events: a systematic review and meta-analysis. JAMA 2012;308:1024−33.

[17] Wu JH, Mozaffarian D. Omega-3 fatty acids, atherosclerosis progression and cardiovascular outcomes in recent trials: new pieces in a complex puzzle. Heart 2014;100:530−3.

[18] Bergkvist C, Berglund M, Glynn A, Wolk A, Akesson A. Dietary exposure to polychlorinated biphenyls and risk of myocardial infarction—a population-based prospective cohort study. Int J Cardiol 2015;183:242−8.

[19] Donat-Vargas C, Gea A, Sayon-Orea C, de la Fuente-Arrillaga C, Martinez-Gonzalez MA, Bes-Rastrollo M. Association between dietary intake of polychlorinated biphenyls and the incidence of hypertension in a Spanish cohort: the Seguimiento Universidad de Navarra project. Hypertension 2015;65:714−21.

[20] Groth E. Scientific foundations of fish-consumption advice for pregnant women: epidemiological evidence, benefit-risk modeling, and an integrated approach. Environ Res 2016;152:386−406.

[21] Mozaffarian D, Micha R, Wallace S. Effects on coronary heart disease of increasing polyunsaturated fat in place of saturated fat: a systematic review and meta-analysis of randomized controlled trials. PLoS Med 2010;7:e1000252.

[22] Farvid MS, Ding M, Pan A, Sun Q, Chiuve SE, Steffen LM, et al. Dietary linoleic acid and risk of coronary heart disease: a systematic review and meta-analysis of prospective cohort studies. Circulation 2014;130:1568−78.

[23] Kagan HB. Victor Grignard and Paul Sabatier: two showcase laureates of the Nobel Prize for Chemistry. Angew Chem Int Ed Engl 2012;51:7376−82.

[24] Schleifer D. The perfect solution. How trans fats became the healthy replacement for saturated fats. Technol Cult 2012;53:94−119.

[25] Willett WC, Ascherio A. Trans fatty acids: are the effects only marginal? Am J Public Health 1994;84:722−4.

[26] Larque E, Zamora S, Gil A. Dietary trans fatty acids in early life: a review. Early Hum Dev 2001;65(Suppl):S31−41.

[27] Mente A, de Koning L, Shannon HS, Anand SS. A systematic review of the evidence supporting a causal link between dietary factors and coronary heart disease. Arch Intern Med 2009;169:659−69.

[28] Howard BV, Manson JE, Stefanick ML, Beresford SA, Frank G, Jones B, et al. Low-fat dietary pattern and weight change over 7 years: the Women's Health Initiative Dietary Modification Trial. JAMA 2006;295:39−49.

[29] Micha R, Mozaffarian D. Trans fatty acids: effects on metabolic syndrome, heart disease and diabetes. Nat Rev Endocrinol 2009;5:335−44.

[30] Forouhi NG, Koulman A, Sharp SJ, Imamura F, Kroger J, Schulze MB, et al. Differences in the prospective association between individual plasma phospholipid saturated fatty acids and incident type 2 diabetes: the EPIC-InterAct case-cohort study. Lancet. Diabetes Endocrinol 2014;2:810−18.

[31] Malik VS, Chiuve SE, Campos H, Rimm EB, Mozaffarian D, Hu FB, et al. Circulating very-long-chain saturated fatty acids and incident coronary heart disease in US men and women. Circulation 2015;132:260−8.

[32] Mensink RP, Zock PL, Kester AD, Katan MB. Effects of dietary fatty acids and carbohydrates on the ratio of serum total to HDL cholesterol and on serum lipids and apolipoproteins: a meta-analysis of 60 controlled trials. Am J Clin Nutr 2003;77:1146−55.

[33] Volk BM, Kunces LJ, Freidenreich DJ, Kupchak BR, Saenz C, Artistizabal JC, et al. Effects of step-wise increases in dietary carbohydrate on circulating saturated Fatty acids and palmitoleic acid in adults with metabolic syndrome. PLoS One 2014;9:e113605.

[34] Chowdhury R, Warnakula S, Kunutsor S, Crowe F, Ward HA, Johnson L, et al. Association of dietary, circulating, and supplement fatty acids with coronary risk: a systematic review and meta-analysis. Ann Intern Med 2014;160:398−406.

[35] Siri-Tarino PW, Sun Q, Hu FB, Krauss RM. Meta-analysis of prospective cohort studies evaluating the association of saturated fat with cardiovascular disease. Am J Clin Nutr 2010;91:535−46.

[36] Abete I, Romaguera D, Vieira AR, Lopez de Munain A, Norat T. Association between total, processed, red and white meat consumption and all-cause, CVD and IHD mortality: a meta-analysis of cohort studies. Br J Nutr 2014;112:762−75.

[37] Thorning TK, Raziani F, Bendsen NT, Astrup A, Tholstrup T, Raben A. Diets with high-fat cheese, high-fat meat, or carbohydrate on cardiovascular risk markers in overweight postmenopausal women: a randomized crossover trial. Am J Clin Nutr 2015;102: 573−81.

[38] Bao Y, Han J, Hu FB, Giovannucci EL, Stampfer MJ, Willett WC, et al. Association of nut consumption with total and cause-specific mortality. N Engl J Med 2013;369: 2001−11.

[39] Rohrmann S, Overvad K, Bueno-de-Mesquita HB, Jakobsen MU, Egeberg R, Tjonneland A, et al. Meat consumption and mortality—results from the European Prospective Investigation into Cancer and Nutrition. BMC Med 2013;11:63.

[40] Babio N, Sorli M, Bullo M, Basora J, Ibarrola-Jurado N, Fernandez-Ballart J, et al. Association between red meat consumption and metabolic syndrome in a Mediterranean population at high cardiovascular risk: cross-sectional and 1-year follow-up assessment. Nutr Metab Cardiovasc Dis 2012;22:200−7.

[41] Feskens EJ, Sluik D, van Woudenbergh GJ. Meat consumption, diabetes, and its complications. Curr Diab Rep 2013;13:298−306.

[42] Li Y, Hruby A, Bernstein AM, Ley SH, Wang DD, Chiuve SE, et al. Saturated fats compared with unsaturated fats and sources of carbohydrates in relation to risk of coronary heart disease: a prospective cohort study. J Am Coll Cardiol 2015;66:1538−48.

[43] Jacobs Jr. DR, Tapsell LC. Food, not nutrients, is the fundamental unit in nutrition. Nutr Rev 2007;65:439−50.

[44] Rissanen A, Hakala P, Lissner L, Mattlar CE, Koskenvuo M, Ronnemaa T. Acquired preference especially for dietary fat and obesity: a study of weight-discordant monozygotic twin pairs. Int J Obes Relat Metab Disord 2002;26:973−7.

[45] Mozaffarian D, Ludwig DS. Dietary guidelines in the 21st century—a time for food. JAMA 2010;304:681−2.

[46] An evaluation of dietary guidance graphic alternatives: the evolution of the eating right pyramid. Nutr Rev 1992;50:275−82.

[47] Mozaffarian D, Hao T, Rimm EB, Willett WC, Hu FB. Changes in diet and lifestyle and long-term weight gain in women and men. N Engl J Med 2011;364:2392−404.

[48] Smith JD, Hou T, Ludwig DS, Rimm EB, Willett WC, Hu FB, et al. Changes in intake of protein foods, carbohydrate amount and quality, and long-term weight change: results from 3 prospective cohorts. Am J Clin Nutr 2015;101:1216−24.

[49]   Alhazmi A, Stojanovski E, McEvoy M, Garg ML. Macronutrient intakes and development of type 2 diabetes: a systematic review and meta-analysis of cohort studies. J Am Coll Nutr 2012;31:243−58.

[50]   Tinker LF, Bonds DE, Margolis KL, Manson JE, Howard BV, Larson J, et al. Women's Health Initiative. Low-fat dietary pattern and risk of treated diabetes mellitus in postmeno-pausal women: the Women's Health Initiative randomized controlled dietary modification trial. Arch Intern Med 2008;168:1500−11.

[51]   Micha R, Mozaffarian D. Saturated fat and cardiometabolic risk factors, coronary heart disease, stroke, and diabetes: a fresh look at the evidence. Lipids 2010;45:893−905.

[52]   Appel LJ, Sacks FM, Carey VJ, Obarzanek E, Swain JF, Miller 3rd ER, et al. OmniHeart Collaborative Research Group. Effects of protein, monounsaturated fat, and carbohydrate intake on blood pressure and serum lipids: results of the OmniHeart randomized trial. JAMA 2005;294:2455−64.

[53]   Gadgil MD, Appel LJ, Yeung E, Anderson CA, Sacks FM, Miller 3rd ER. The effects of carbohydrate, unsaturated fat, and protein intake on measures of insulin sensitivity: results from the OmniHeart trial. Diabetes Care 2013;36:1132−7.

[54]   Martinez-Gonzalez MA, Bes-Rastrollo M. Dietary patterns, Mediterranean diet, and cardiovascular disease. Curr Opin Lipidol 2014;25:20−6.

[55]   Mozaffarian D, Ludwig DS. The 2015 US Dietary Guidelines: lifting the ban on total dietary fat. JAMA 2015;313:2421−2.

Chapter 4

Virgin Olive Oil: A Mediterranean Diet Essential

Almudena Sánchez-Villegas and Ana Sánchez-Tainta

4.1 INTRODUCTION

Olive oil has been recognized for centuries for its nutritional properties and considered as the "elixir of youth and health" by antique Greeks. Olive oil is the main fat source used within the Mediterranean countries. In fact, olive oil is a key element in the traditional Mediterranean Diet, being the main source of fat within the pattern. Thus, the high content in fat of the Mediterranean Diet (up to 40% of total energy) due to this element is one of the most important distinguishing marks of this ancestral dietary pattern. In this context, it has been accepted that olive oil (particularly virgin olive oil) plays a crucial role in the cardioprotective effect of the Mediterranean Diet, attributed to its beneficial effects on many cardiovascular risk factors including lipid profiles, inflammation, endothelial function, blood pressure, insulin resistance, or weight control.

4.2 COMPOSITION OF OLIVE OIL

The major components of olive oil are fatty acids that represent more than 98% of its composition. They are mainly monounsaturated fatty acids (MUFAs) such as oleic acid (55%−85% of the total fatty acids), although olive oil also contains polyunsaturated fatty acids (PUFAs) such as linoleic acid and saturated fatty acids (SFAs) like stearic or palmitic acids. Among the minor components that constitute from 1% to 2% of total content of an olive oil there are two types: (1) the unsaponifiable fraction that includes aliphatic and triterpene alcohols, sterols, hydrocarbons such as squalene, volatile compounds, tocopherols, and carotenes and (2) the soluble fraction that includes the phenolic compounds. A key olive oil polyphenol is oleuropein, which accounts for approximately 80% of olive oil phenolic content. Other important phenolic compounds are phenolic alcohols (hydroxityrosol and tyroxol generated from oleuropein), flavonoids, lignans, and secoiridoids (Table 4.1). The minor components of olive oil are of special importance in

The Prevention of Cardiovascular Disease Through the Mediterranean Diet.
DOI: http://dx.doi.org/10.1016/B978-0-12-811259-5.00004-4

TABLE 4.1 Composition of Olive Oil

Saponifiable Fraction	Insaponifiable Fraction	Phenolic Compounds
Oleic acid (55%−85%)	Hydrocarbons (squalene)	Simple phenols (hydroxityrosol and tyroxol)
	Volatile compounds	
Linoleic acid (3%−21%)	Tocopherols	Flavonoids
Linolenic acid (max. 1%)	Sterols	Lignans
Palmitic acid (7.5%−20%)	Carotenes	Secoiridoids (oleuropein)
Stearic acid (0.5%−5%)	Aliphatic and triterpene alcohols	
	Pigments (chlorophyll)	

cardiovascular health because their antiinflammatory and antioxidant effects and because their properties that favorably modulate hemostatic factors and improve the stability of arteriosclerotic plaque.

4.3 NATURAL JUICE OF THE OLIVE, WHY EXTRA-VIRGIN?

The olive tree *Olea europea* L. is one of the oldest agricultural tree crops and source of olive oil and olives. Agronomic and technological aspects of the olive oil production such as the processing system, ripeness of the olives at harvesting, climate, or cultivar characteristics like use of irrigation have a direct impact in the concentration of minor components of olive oil. Only oils obtained by mechanical extraction and that have not undergone any treatment other than washing, decantation, centrifugation, or filtration are called virgin olive oils. Extra-virgin oil would have even further quality assessment. Oils differ in bioactive composition depending on refining process. Specifically, phenolic compounds and to lesser degree squalene are lost during refining process and only are present in virgin and extra-virgin olive oil.

So, not all olive oils that are in the market are equal. Depending on the manufacturing process, we can found different types of oil:

- **Virgin olive oil**: obtained directly from the ripe fruit by mechanical procedures. It is the only consumed raw without the use of any solvents, so it keeps all its properties intact. Olive oil has excellent antioxidants such as polyphenols and vitamin E, which are lost if the oil is subjected to refining processes. "Extra-Virgin" olive oils are those that have no taste defects and have a very low acidity rate (<0.8%). They are the most expensive ones. Those labeled as "Virgin" olive oils have modest taste defects and a slightly higher acidity level (<2%).

- **Pure olive oil**: which is a *blend* of refined olive oil and virgin olive oil. In the market, there are varieties of intense flavor and mild taste depending on whether more or less virgin oil olive is added, respectively.
- **Olive pomace oil**: which is obtained from the pulp and seeds of olives after extraction of virgin olive oil using chemical solvents. It must pass a refining process and virgin olive oil is added after to make it suitable for consumption.

4.4   EFFECTS ON HEALTH

Olive oil consumption has been associated with a lower risk of cardiovascular disease (that includes coronary heart disease, stroke, and peripheral artery disease) and with longevity and also with a lower risk of other diseases such as diabetes, hypertension, obesity, cancer, neurodegenerative diseases, and depression.

In fact, in 2004, the Federal Drug Administration of the United States permitted a claim on olive oil labels concerning: "the benefits on the risk of coronary heart disease of eating about two tablespoons (23 g) of olive oil daily, due to the MUFA in olive oil."

Mechanisms from which olive oil could exert its function are the improvement of lipid profile, insulin sensitivity, coagulation, blood pressure and endothelial function, and the reduction in oxidation susceptibility of LDL cholesterol and in the low-grade inflammation (Table 4.2).

4.4.1   Effects on Triglycerides, LDL Cholesterol, and HDL Cholesterol

LDL and HDL cholesterol are lipoproteins implicated in the etiopathogenesis of the cardiovascular disease. Whereas LDL particles carry about two-thirds of plasma cholesterol and can infiltrate the arterial wall leading to atherosclerosis, HDL particles are antiatherogenic as their role is to deliver cholesterol to the liver to be metabolized and eliminated. Moreover, HDL is able to dislodge cholesterol from the arterial walls.

Traditionally, the nutritional recommendation has been to reduce total fat intake and to substitute its energy fraction by calories coming from carbohydrates. Nevertheless, this recommendation has been questioned because carbohydrates when replaced to fats not only are able to reduce LDL levels in a beneficial way but also HDL levels in a detrimental one and to increase triglycerides levels [1]. So, the key message should be centered in replacing some types of fats by others.

Although the results form a meta-analysis based on 14 studies found that the replacement of SFA (such as butter) by oils enriched in MUFA (such as

TABLE 4.2 Mechanisms From Which Extra-Virgin Olive Oil Could Exert Its Cardioprotective Function

Blood lipids	
HDL cholesterol	↑
LDL cholesterol	↓
Total cholesterol	↓
Triglycerides	↓
LDL particle size	↑
Insulin resistance	↓
Blood pressure	↓
Endothelial function	↓ cell adhesion molecules (VCAM-1, ICAM-1)
Low-grade inflammation	↓ inflammatory markers (IL-6, C-reactive protein, MCP-1)
Oxidative stress	
LDL oxidation	↓
HDL oxidation	↓
Hemostasis	↓ coagulation factors and platelet aggregation

olive oil) or PUFA (such as other vegetal oils) has similar effects on total, LDL and HDL cholesterol levels, there was an important difference, the lipoprotein oxidation [2]. In this sense, it's worth noting that oxidized LDLs are important risk markers of cardiovascular disease. In fact, they have been shown to be independently associated with 10-year coronary artery disease events in general population [3]. In addition, HDL oxidation reduces HDL functionality.

Diets rich in oleic acid are able to reduce the susceptibility of LDL to oxidation in a higher degree than diets enriched in other oils. Olive oil consumption has been also associated with lower HDL oxidation. Moreover, it has to be highlighted that LDL particle size is relevant since small-size and denser particles are more prone to oxidation and can better enter into the arterial wall as compared to larger LDL particles. In fact, smaller, denser LDL particles have been associated with an increased risk for coronary heart disease in prospective studies. HDL cholesterol particles are also heterogeneous and evidence suggests that larger diameter HDL particles may be more cardioprotective. The fat content of the diet influences the LDL and HDL particle size. In fact, low-fat diets led to a decrease in the size whereas high-MUFA diets (such as those including olive oil like the Mediterranean Diet) lead to larger LDL particles.

However, although MUFA and particularly oleic acid may exert beneficial effects in lipid profile and its oxidation, evidences have accumulated on the beneficial properties of minor though highly bioactive components of olive oil (virgin and extra-virgin). Most of the intervention trials with olive oil have investigated the effect of phenolic compounds in the prevention of oxidation of LDL and HDL particles. LDL oxidative process is an initiating factor for atherosclerotic plaques. Thus, the antioxidant effect of phenolic compounds can prevent lipid peroxidation and consequently, oxidative modification of LDL and, finally, the initiation of atherosclerotic processes. One of the most important findings comes from the EUROLIVE (the effect of olive oil consumption on oxidative damage in European populations) study that provided clear evidence that virgin olive oil is more than just MUFA. The EUROLIVE dietary intervention study is a multicenter clinical trial performed in 200 individuals from five European countries. Participants were assigned to receive 25 ml/d of three similar olive oils only differing in their phenolic content (from 2.7 to 366 mg/kg of olive oil). This study showed that *in vivo* consumption of olive oil with three different phenolic concentrations increased HDL cholesterol and decreased total and LDL-cholesterol and triglycerides in a dose-dependent manner. Moreover, oxidative stress markers decreased linearly with increasing phenolic content [4,5].

Also the PREDIMED trial found similar results. The PREDIMED trial (PREvención con DIeta MEDiterránea in Spanish) is a nutritional intervention study composed of 7447 participants at high risk for cardiovascular disease. The participants of the PREDIMED trial were assigned to three different intervention arms, two of the groups received educational counseling regarding Mediterranean Diet and a supplementation with extra-virgin olive oil or with nuts whereas the third group was endorsed to follow a low-fat diet recommended by the American Heart Association. In this trial, participants assigned to Mediterranean Diet supplemented with extra-virgin olive oil (with high content in polyphenols) showed a significant decrease in LDL oxidation compared to the control group, confirming previously obtained results [6,7].

4.4.2 Antioxidant and Antiinflammatory Effect

Cellular oxidative stress is mediated through mitochondrial membrane oxidative injury. Mitochondrial membranes are very sensitive to free radicals attack because the presence of double bond carbon-carbon in their structure. Thus, a low level of fatty acid unsaturation (as that of the oleic acid) might lead to a decrease cellular oxidative stress [8]. However, most of the benefits attributed to olive oil and related to its antiinflammatory and antioxidant capacity are exerted by its minor components. For example, although the quantities of α-tocopherol (vitamin E) and carotenoids present in a daily consumption of virgin olive oil are low, its chronic ingestion contributes to the overall pool of antioxidants in the human body. Also tyrosol, hydroxytyrosol, and lignans have shown antioxidant properties in experimental studies. Triterpenes have shown antiinflammatory and antioxidant properties in *in vitro* studies.

Chronic inflammation is a critical factor in the pathogenesis of many inflammatory disease states including cardiovascular disease, cancer, diabetes, degenerative joint diseases, and neurodegenerative diseases. An important number of polyphenolic compounds from virgin olive oil have antiinflammatory properties. One of these compounds is the oleocanthal. The antiinflammatory properties of this compound have been compared to those exerted by ibuprofen. Several studies have found that the consumption of virgin olive oil with high phenolic content is able to reduce inflammatory markers such as interleukin-6 (IL-6) or C-reactive protein. In fact, an analysis within the PREDIMED trial found a significant reduction in C-reactive protein among participants assigned to follow a Mediterranean dietary pattern supplemented with extra-virgin olive oil [9]. In this trial, also urinary excretion of polyphenols was measured. Participants with the highest increase in urinary polyphenol excretion during follow-up period showed significant lower plasma inflammatory biomarkers IL-6, tumor necrosis factor-α (TNF-α), and monocyte chemotactic protein-1 (MCP-1) than those with the lowest level of excretion [10].

A recent meta-analysis has synthesized data from randomized controlled trials investigating the effect of olive oil on markers of inflammation. The authors conclude that olive oil interventions (with daily consumption ranging approximately between 1 and 50 mg) resulted in a significantly more pronounced decrease in C-reactive protein and IL-6 as compared to controls. However, the authors also state that due to the heterogeneous study designs (e.g., olive oil given as a supplement or as part of a dietary pattern, variations in control diets), a conservative interpretation of the results is necessary [11].

4.4.3   Effect on Endothelial Function

Low-grade chronic synthesis and release of proinflammatory cytokines such as IL-1 (interleukin-1) or TNF-α within the vascular wall affect endothelial function via, for example, upregulation of adhesion molecules. The endothelium is a vascular tissue that lines the interior surface of blood vessels. Endothelial dysfunction is the first pathological symptom of anatomical lesions in atherosclerosis process. This dysfunction implies a breakdown in the defense mechanism of the endothelial wall, which in turn induces a prothrombotic state, activates the inflammatory process, and alters the vasomotor regulation of the vascular wall. This is why the endothelium and the factors, which can damage it, are now considered so important. Endothelial dysfunction and an excess of endothelial cell adhesion molecules are predictors of both pathogenic mechanisms of atherosclerosis and cardiovascular disease.

In experimental studies, the minor components of olive oil have been described as beneficial factors to improve endothelial function by decreasing the expression of cell adhesion molecules. However, the effect of phenolic compounds on cell adhesion molecules is controversial. Some studies have found a decrease in cell adhesion molecules after intake of phenol-rich virgin olive oil compared with refined olive oil [12], whereas other studies have failed to report a significant association between the consumption of virgin olive oil with high content of phenolic compounds and VCAM-1 (vascular cell adhesion molecule-1) and ICAM-1 (intercellular adhesion molecule-1) plasma concentrations [13].

In a subsample within the PREDIMED trial, Mena et al. found a decrease in VCAM-1 among the individuals assigned to the Mediterranean Diet supplemented with extra-virgin olive oil [14]. Also participants with the higher increase in polyphenols excretion during follow-up showed significant lower plasma VCAM-1 and ICAM-1 in a subanalysis within the trial [10].

4.4.4 Antithrombotic Effect

Hemostasis is a fundamental process in the maintenance of circulation and is the result of a complex equilibrium between coagulation and fibrinolysis. Thrombogenesis is a complex process, which involves the platelets activation (primary hemostasis), the mechanisms of coagulation (secondary hemostasis) and fibrinolysis, a system that has been implicated in the reabsorption of thrombus.

An accrual of studies have analyzed the role of olive oil consumption in reduction of thrombogenesis through a decrease in coagulation factors and platelet aggregation [15]. MUFA have positive effects on a number of factors responsible for hemostasis such as platelet aggregation and factor VII (protein that initiates the process of coagulation). Moreover, MUFA are able to decrease the plasmatic levels of Von Willebrand factor, a fundamental component of platelet adhesion and aggregation processes. Other proteins implicated in hemostasis are the plasminogen activator inhibitor-1 (PAI-1) and the tissue factor pathway inhibitor (TFPI). Several intervention studies have observed a reduction in plasmatic levels of these proteins after MUFA or oleic acid intake. Apart from MUFA, other minor components of olive oil also influence markers on hemostasis. Polyphenols are able to inhibit platelet-inducted aggregation and to show antithrombotic properties in both experimental and human intervention studies.

4.5 EPIDEMIOLOGICAL EVIDENCES REGARDING THE ROLE OF OLIVE OIL IN CARDIOVASCULAR DISEASE

In the 1950s of the 20th century the pioneer Seven Countries Study conducted by Ancel Keys, followed by the MONICA study (the MONICA Project (MONItoring CArdiovascular disease) was conceived in 1979, in which teams from 38 populations in 21 countries studied heart disease, stroke, and risk factors from the mid-1980s to the mid-1990s), and other ecological studies observed lower mortality rates in Mediterranean countries than in other European countries or in the United States, in spite of their high fat intake [16,17]. Moreover, incidence and mortality of coronary heart disease was lower among these countries. The type of fats used to cook differed substantially from one to other country with the use of margarine and butter in the northern Europe and the USA and the use of olive oil for cooking and dressing salads in the Mediterranean regions. However, although olive oil is believed to play a key role on these health benefits attributed to the Mediterranean dietary pattern and considerable evidence on the biological mechanisms involved has been published there is relatively little direct epidemiological evidence on the individual effects of olive oil on cardiovascular disease or on risk factors of cardiovascular disease (Table 4.3).

With regard to case-control and cohort studies, only a few of them have analyzed the association between olive oil consumption and the risk of coronary heart disease and/or stroke with inconsistences in outcomes between studies.

A Spanish case-control study composed by 171 cases and the same number of controls found that those participants who referred a high intake of olive oil showed an important reduction (around 80%) in the risk of a first myocardial infarction [18].

Likewise, other similar study, but in Greece and based on 848 cases and 1078 controls, concluded that the exclusive consumption of olive oil was able to reduce the risk of any acute coronary syndrome in a 47% independently of weight, smoking, physical activity level, educational status, the presence of family history of coronary heart disease, as well as hypertension, hypercholesterolemia, and diabetes. On the other hand, consumption of olive oil in combination with other oils or fats was not significantly associated with lower risk of acute coronary syndrome compared to no olive oil consumption [19].

However, Bertuzzi et al. analyzed the relation between olive oil consumption and the risk of nonfatal myocardial infarction in an Italian case-control study. The authors did not find a strong relation between olive oil and risk of this coronary event [20]. Neither found Gramenzi et al. a significant association in a sample of 287 women who had had an acute myocardial infarction (median age 49, range 22−69 years) and 649 controls with acute disorders unrelated to ischemic heart disease [21].

TABLE 4.3 Epidemiological Evidences Regarding the Association Between Olive Oil Consumption and Cardiovascular Disease

Reference	Study	Design	Exposure	Outcome	Main Results
Keys et al. Am J Epidemiol 1986; 124: 903–15	Seven Countries	Ecological	Total diet	Death rates	Lower rates of all-cause and coronary heart disease in cohorts with olive oil as main fat
Fernández Jarne et al. Int J Epidemiol 2002; 31: 474–80		Case-control	Olive oil	Myocardial infarction	Reduction 80%
Kontogianni et al. Clin Cardiol 2007; 30: 125–9	CARDIO2000	Case-control	Olive oil	Acute coronary syndrome	Reduction 47%
Bertuzzi et al. Int J Epidemiol 2002; 31: 1274–7		Case-control	Olive oil	Nonfatal myocardial infarction	No relation
Gramenzi et al. BMJ 1990; 300: 771–3		Case-control	Olive oil	Myocardial infarction	No relation
Buckland et al. Am J Clin Nutr 2012; 96: 142–9	EPIC-Spain	Prospective cohort	Olive oil	Cardiovascular mortality	Reduction 44%
Buckland et al. Br J Nutr 2012; 108: 2075–82	EPIC-Spain	Prospective cohort	Olive oil	Coronary heart disease	Reduction 22%
Dilis et al. Br J Nutr 2012; 108: 699–709	EPIC-Greece	Prospective cohort	Olive oil	Stroke	Reduction 20%

(Continued)

TABLE 4.3 (Continued)

Reference	Study	Design	Exposure	Outcome	Main Results
Misirli et al. Am J Epidemiol 2012; 176: 1185–92	EPIC-Greece	Prospective cohort	Olive oil	Stroke mortality	Reduction 11%
Samieri et al. Neurology 2011; 77:418–25	Three-City Study	Prospective cohort	Olive oil	Stroke	Reduction 41%
Bendinelli et al. Am J Clin Nutr 2011; 93: 275–83	EPICOR (EPIC-Italy)	Prospective cohort	Olive oil	Coronary events	Reduction 60%
Guasch-Ferré et al. BMC Med 2014; 12: 78	PREDIMED	RCT (analyzed as an observational cohort)	Olive oil and extra-virgin olive oil	Cardiovascular disease	Olive oil 35% reduction, extra-virgin olive oil 39% reduction
Barzi et al. Eur J Clin Nutr 2003; 57: 604–11	GISSI-Prevention trial	Secondary prevention trial	Olive oil	Overall mortality	Reduction 24%
Estruch et al. N Engl J Med 2013; 368: 1279–90	PREDIMED	RCT	Extra-virgin olive oil	Cardiovascular disease	Reduction 30%
Ruiz-Canela et al. JAMA 2014; 311:415–17	PREDIMED	RCT	Extra-virgin olive oil	Peripheral artery disease	Reduction 66%
Martínez-González et al. Circulation 2014; 130:18–26	PREDIMED	RCT	Extra-virgin olive oil	Atrial fibrillation	Reduction 40%

Martínez-González et al. Br J Nutr 2014; 112: 248–59	Meta-analysis	Olive oil	Coronary heart disease and stroke	Reduction 18%
Schwingshackl et al. Lipids Health Dis 2014; 13: 15	Systematic review	MUFA, olive oil, oleic acid and MUFA/SFA	Cardiovascular mortality and Cardiovascular events	30% reduction with olive oil

MUFA: Monounsaturated fatty acids; SFA: Saturated fatty acids.

Within the observational prospective studies it is worth to point out the EPIC study (European Prospective Investigation into Cancer and Nutrition). The EPIC study is a prospective cohort study developed in ten European countries and designed to assess the association between diet and cancer risk. However, other events have been collected and analyzed including coronary heart disease, stroke, or overall cardiovascular incidence and mortality. Both EPIC-Spain and EPIC-Greece have analyzed the effect of olive oil consumption on cardiovascular risk and specifically on coronary risk. In 2012, Buckland et al. found an inverse relationship between olive oil consumption and the risk of cardiovascular mortality within the EPIC-Spain cohort [22]. After following up to 40,622 participants (62% female) aged 29−69 years from five Spanish regions in 1992−1996, they collected 416 cardiovascular deaths. Participants with a high consumption of olive oil (30 or more grams per day) as compared to those with a low intake (<14 g/d) had a significant risk reduction (44%) in cardiovascular mortality. Moreover, the researchers found a dose−response relationship with higher effects when the consumption of the food item was increased.

Also in 2012, these authors published new results ascertaining, specifically, the risk of coronary heart disease in the same sample. In this case, the authors also found a 22% risk reduction for coronary heart disease associated to elevated consumption of olive oil, although, in this case, the association was not statistically significant. It is worth to highlight that a difference in the risk of coronary heart disease associated to the type of consumed olive oil was found. Those participants who consumed virgin olive oil showed a higher reduction in the risk of coronary events than those consuming refined olive oil although again, the difference did not reach the statistical significance [23].

The results from the EPIC-Spain contrast with those reported in the EPIC-Greek cohort, which found that olive oil was not associated with a reduction of the coronary heart disease risk or mortality [24]. However, in this sample, when stroke was analyzed, the researchers found a beneficial effect for olive oil with a significant reduction of 20% in the risk. Nevertheless, the reduction in stroke mortality was lower and no statistically significant (11% of reduction) [25].

In the Three-city Study in France, the researchers also analyzed the association between the consumption of olive oil and the risk of stroke. After adjustment for sociodemographic and dietary variables, physical activity, body mass index, and risk factors for stroke, a lower incidence for stroke with higher olive oil use was observed. Compared to those who never used olive oil, those with intensive use (participants who used olive oil for cooking and dressing) had a 41% lower risk of stroke [26].

Other remarkable study is the EPICOR study in Italy. The EPICOR study is a collaborative prospective investigation that aims to estimate the risk of cardiovascular diseases associated with dietary and lifestyle habits in the Italian cohort of the EPIC Study. In an analysis published in 2011 and based entirely on women, the researchers observed that those women with the highest consumption of olive oil (more than 31 g per day) as compared to those with the lowest consumption (less than 16 g) showed a reduction in the risk of coronary events in the overall sample (approximately 30,000 women) and in an specific analysis with postmenopausal women (45% of the sample). In both cases, the reduction in the risk was around 60% associated to an elevated consumption of olive oil. Moreover, a dose−response relationship was described with the highest benefits referred for the highest consumption [27].

One of the most recent findings come from the PREDIMED trial. Although the PREDIMED trial is a controlled study, some analyses are conducted treating the sample as an observational prospective cohort study. In this sense, in 2014, the PREDIMED team found that participants in the highest level of total olive oil and extra-virgin olive oil consumption had 35% and 39% cardiovascular disease risk reduction, respectively, compared to the reference with the lowest consumption. Higher baseline total olive oil consumption was also associated with 48% (reduced risk of cardiovascular mortality). For each 10 g/d increase in extra-virgin olive oil consumption, cardiovascular disease and mortality risk decreased by 10% and 7%, respectively. To the contrary, consumption of common olive oil was not significantly associated with cardiovascular morbidity and mortality [28].

Some studies have analyzed the role of olive oil in overall mortality in participants who had suffered a prior myocardial infarction. Barzi et al., for example, in the GISSI-Prevenzione trial found that olive oil consumption among these patients was able to reduce their mortality by 24% [29].

With regard to intervention (experimental) studies in humans is worth to mention the Lyon Diet Heart Study and the PREDIMED trial. The Lyon Diet Heart Study was a randomized, controlled trial designed to test the effectiveness of a Mediterranean-type diet on composite measures of the coronary recurrence rate after a first myocardial infarction. Subjects in the experimental group were instructed to adopt a Mediterranean-type diet. However, the intervention was not centered in the use of olive oil. The researchers supplied the participants with a margarine (to replace butter and cream) with a content of SFA (15% kcal) and oleic acid (48% kcal but 5.4% kcal 18:1 trans) comparable to that found in olive oil, with the exception that the margarine was higher in linoleic acid (16.4% versus 8.6% kcal) and more so in α-linolenic acid (4.8% vs. 0.6% kcal). Exclusive use of rapeseed oil and olive oil was recommended for salads and food preparation. Use of olive oil exclusively was not recommended because it was not acceptable as the only oil source in the diet. The study found that participants in the intervention group (following a Mediterranean Diet supplemented with this margarine) showed a 50%−70% lower risk of recurrent heart disease [30].

To our knowledge, only one controlled trial has evaluated the role of olive oil consumption in cardiovascular risk. The PREDIMED trial found an important reduction in the risk of cardiovascular disease among those participants assigned to receive educational counseling regarding Mediterranean Diet and a supplementation with extra-virgin olive oil. The hazard ratio (HR) was 0.70 (95% confidence interval [CI]: 0.54−0.92). So, a significant reduction of 30% in cardiovascular risk (a composite of myocardial infarction, stroke, and death from cardiovascular causes) was observed for this intervention as compared to those participants assigned to a control low-fat diet [31]. A protective effect for extra-virgin olive oil was also observed for peripheral artery disease [32] and for atrial fibrillation (arrhythmia) [33] in the PREDIMED trial. The risk reduction for peripheral arterial disease was even stronger, 66% in the group assigned to the Mediterranean Diet supplemented with extra-virgin olive oil as compared to the group assigned to the control diet. In addition, whereas for the analyses of cardiovascular risk and peripheral artery disease not only olive oil but also nut consumption were associated to risk reduction, the relative risk reduction against atrial fibrillation within the PREDIMED trial was observed only for the intervention with extra-virgin olive oil (not with the intervention with nuts) with a significant reduction near 40%. One of the explanations argues by the researchers was the antiinflammatory and antioxidant effects of extra-virgin olive oil attributed to its richness in phenolic compounds.

All these results have been included in several systematic reviews and meta-analyses such as that published by Martinez-Gonzalez et al. in the *British Journal of Nutrition* in 2014. The authors found a significant reduction (around 18%) in both coronary heart disease and stroke (jointly analyzed) associated to a 25 g increase in olive oil consumption, more important for stroke than for coronary heart disease [34]. When only cohort studies were analyzed, a significant reduction near 25% was found for stroke and only 4% and no significant risk reduction associated to 25 g increase in olive oil for coronary heart disease. Although the inverse association between olive oil and risk of stroke was consistent, it's important to highlight the scarcity of the studies on this outcome. On the other hand, this review points out toward a lack of effect of olive oil on coronary heart disease risk. The authors argue several reasons to explain this apparent paradox: (1) several bias regarding exposure assessment and analyses should be taken into account; (2) maybe any single dietary factor is unlikely to have a large effect on the risk of the disease; and (3) the appropriate exposure to be assessed should probably be virgin olive oil instead of all kinds of olive oils mixed together. In this sense, as we have mentioned repeatedly the effect of minor polyphenolic components of olive oil which are only present in the virgin variety of this oil but not in refined variety would be able to make a large difference in the potential protection that olive oil could exert on coronary heart disease risk.

In this sense, Schwingshackl et al. analyzed in other systematic review (including only cohort studies) the role of MUFA, olive oil, oleic acid, and the ratio MUFA/SFA in cardiovascular mortality and cardiovascular events (coronary heart disease plus stroke) comparing the maximum versus the minimum intake of each element. MUFA intake was associated neither with cardiovascular events nor with cardiovascular mortally whereas olive oil consumption was associated with a significant reduction in the risk of cardiovascular events with a reduction of almost 30% [35]. The lack of effect of MUFA in cardiovascular risk has been consistently referred in other systematic reviews such as that carried out by Chowdhury et al. [36]. In fact, the results of a large study analyzing data from several important cohort studies, the "Pooling Project of Cohort Studies on Diet and Coronary Disease" found a potential harmful association between MUFA intake and the risk of coronary heart disease, which would run against the potential beneficial effect of olive oil. Jakobsen et al. observed that the replacement of SFA by MUFA marginally increased the risk of coronary events, whereas no significant effects on coronary death were observed [37]. Thus, at least, these results do not support that olive oil may exert cardiovascular effect mainly because its content in MUFA. A potential explanation could be again the sources of MUFA. Most of the studies included in the Pooling Project were based on US populations where the most important source of MUFA are animal fats and not olive oil. In the Nurses' Health Study, an American prospective cohort study, MUFA intake was highly correlated with SFA intake. In the EPIC cohorts, in general, MUFA intake was higher in the central and northern cohorts. However, predominant sources of MUFA differed between cohorts. Whereas in Greece, Spain, and Italy fat of plant origin (mainly olive oil) provided up to 64% of MUFA intake, in other EPIC centers the main sources were meat, meat products, added fats, and dairy products.

A second probable explanation is that the cardioprotective effect of olive oil is mainly due to its content in minor components such as polyphenols, squalene, or carotenes. In this sense, analyses within the PREDIMED trial reported an inverse association between habitual polyphenol intake and incidence of cardiovascular events [38]. The authors observed a significant reduction of cardiovascular events and cardiovascular mortality associated to a higher intake of total polyphenols, lignans, flavanols, and hydroxybenzoic acids. However, although the potential role of several polyphenols in reducing cardiovascular risk can be explained due to the high consumption of olive oil, as it is the case of lignans, other important sources of polyphenols have to be taken into account such as fruits, vegetables, legumes, or red wine also consumed within this Mediterranean sample.

4.6 EPIDEMIOLOGICAL EVIDENCES REGARDING THE ASSOCIATION BETWEEN OLIVE OIL CONSUMPTION AND CARDIOVASCULAR RISK FACTORS

Several epidemiological studies have studied the association between olive oil consumption and several cardiovascular risk factors (Table 4.4). Most of them were carried out among Mediterranean populations. In this sense, it is worth to mention a Spanish sample composed of 4572 adults, the Di@bet.es. In this cross-sectional study, the consumption of olive oil (vs. sunflower oil) was associated to a lower prevalence of obesity, impaired glucose regulation, hypertriglyceridemia, and with low probability of having low HDL cholesterol levels [39].

Some cohort studies such as the SUN cohort study (Seguimiento Universidad de Navarra in Spanish) or the EPIC have also reported interesting results regarding the possible role of olive oil in several risk factors for cardiovascular disease. However, maybe the most representative study that has analyzed the role of olive oil consumption in cardiovascular risk factors is the PREDIMED trial, as we have already mentioned. In this trial, those assigned to receive educational counseling to follow a Mediterranean Diet and a supplementation with extra-virgin olive oil showed a reduction in most of the cardiovascular risk factors with benefits on blood pressure, glycemic control in diabetics, endothelial function, oxidative stress and lipid profiles (decreasing tryglycerides, increasing HDL- and lowering total and LDL cholesterol) and reduces susceptibility of LDL to oxidation and concentrations of inflammatory markers such as C-reactive protein and IL-6. In addition, the olive oil-rich diet was effective in the prevention of diabetes, the metabolic syndrome, and weight gain. The main results obtained from the cited studies will be explained in the following paragraphs.

4.6.1 Diabetes

Diets high in SFA consistently impair both insulin sensitivity and blood lipids, while substituting MUFA for SFA revers these abnormalities. In fact, some studies have suggested that replacing SFAs with PUFAs or MUFAs has beneficial effect in the prevention of type 2 diabetes mellitus. MUFA intake has been shown to decrease insulin resistance. Moreover, high-MUFA diets have been found as good strategies to improve glycemic control and lipoprotein profile in diabetic patients. Beyond, oleic acid, also bioactive compounds such as squalene, tocopherols and polyphenols could exert by different mechanisms favorable effects on insulin sensitivity and diabetes.

In the United States, to our knowledge, only one analysis merging two prospective cohort studies has analyzed the association of olive oil consumption and the risk of diabetes. The Nurses' Health Study I and Nurses' Health study II, found, after 22 years of follow-up, that those women who consumed

TABLE 4.4 Epidemiological Evidences Regarding the Association Between Olive Oil Consumption and Cardiovascular Risk Factors

Author, Year	Study	Design	Exposure	Risk Factor	Main Results
Guasch-Ferré et al. (2015)	NHS I, NHS II	Cohort	Total, salad dressing and added to food or bread OO	Diabetes	>1 tablespoon (>8 g) of total OO/d, reduction (10%), added to food OO, reduction (15%) No significant reduction for salad dressing
Marí-Sanchis et al. (2011)	SUN	Cohort	OO	Diabetes	No effect
InterAct Consortium et al. (2011)	EPIC-Interact study	Cohort	MedDiet (includes OO)	Diabetes	Reduction (12%) for maximum scoring; no reduction after excluding OO from the score
Salas-Salvadó et al. (2014)	PREDIMED	RCT	MedDiet + EVOO	Diabetes	Reduction (near 40%)
Tresserra-Rimbau et al. (2016)	PREDIMED	RCT	Polyphenols (not specific from OO)	Diabetes	Reduction (28%)
Witteman et al. (1989)	NHS	Cohort	MUFA	HTA/BP	No effect on risk of HT
Stamler et al. (2002)	Chicago Western Electric Company study	Cohort	MUFA	HTA/BP	No effect on risk of HT

(Continued)

TABLE 4.4 (Continued)

Author, Year	Study	Design	Exposure	Risk Factor	Main Results
Psaltopoulou et al. (2004)	EPIC–Greece	Cohort (cross-sectional)	MUFA/SFA ratio OO	BP	Reduction in BP
Alonso et al. (2004)	SUN	Cohort	OO	HTA	Significant reduction
Toledo et al. (2013)	PREDIMED	RCT	MedDiet + EVOO	BP	Significant reduction in diastolic BP
Ferrara et al. (2000)		RCT	EVOO vs. sunflower oil	HTA	Reduced need for antihypertensive medication
Moreno-Luna et al. (2012)		RCT	Polyphenol rich OO as compared to a refined oil	BP	Significant reduction in BP for polyphenol rich OO
Medina-Remón et al. (2015)	PREDIMED	RCT	MedDiet + EVOO Polyphenols excreted in urine	BP	Significant reduction in BP
Bes-Rastrollo et al. (2006)	SUN	Cohort	OO	Weight change Obesity	No effect
Rázquin et al. (2009)	PREDIMED	RCT	Plasmatic antioxidant profile (seen for MedDiet + EVOO)	Weight gain	Reduction

Alvarez et al. (2016)	PREDIMED	RCT	MedDiet + EVOO	Anthropometric and body composition parameters	No effect
Babio et al. (2014)	PREDIMED	RCT	MedDiet + EVOO	Metabolic syndrome	Significant reversion, no effect in incidence
Mayneris-Perxachs et al. (2014)	PREDIMED	RCT	Plasma level of oleic acid	Metabolic syndrome	Reduction in prevalence

NHS: Nurses' Health Study; OO: Olive oil; SUN: Seguimiento Universidad de Navarra; EPIC: European Prospective Investigation into Cancer and Nutrition; PREDIMED: PREvención con DIeta MEDiterránea; RCT: Randomized Controlled Trial; MedDiet: Mediterranean Diet; EVOO: Extra-virgin olive oil; MUFA: Monounsaturated fatty acids; SFA: Saturated fatty acids; HTA: Hypertension; BP: Blood pressure.

>1 tablespoon (>8 g) of total olive oil per day compared to those who never consumed olive oil had a significant reduction in the risk of diabetes (10%). The corresponding risk reductions were 5% (no significant) for salad dressing olive oil and 15% (significant) for olive oil added to food or bread. The authors estimated that substituting olive oil (8 g/d) for stick margarine, butter, or mayonnaise was associated with 5%, 8%, and 15%, respectively, lower risk of diabetes mellitus type 2. The authors of this study highlighted the favorable effects found for olive oil on diabetes prevention despite the relatively low intake of this fat compared with their Mediterranean counterparts [40].

Outside the United States, some studies from Mediterranean countries such as Spain are available. Nevertheless, only a prospective study has analyzed the specific association of olive oil as unique dietary factor with diabetes risk. The analyses carried out within the Seguimiento Universidad de Navarra (SUN) cohort study in Spain. The SUN study is a Spanish prospective cohort study based on university graduates that includes more than 20,500 participants. However, the authors did not find any significant association between olive oil consumption and diabetes incidence. Nonetheless, the small number of cases of diabetes on the sample might have contributed to the null finding [41].

Most of the examples, unfortunately, have analyzed olive oil consumption but in the frame of an overall dietary pattern, the Mediterranean Diet. In this sense, the results from the EPIC-Interact study in an analysis of participants, suggested that olive oil, as one of the key components of the Mediterranean Diet could be responsible for the beneficial effect found for participants with high adherence to a Mediterranean dietary score (MDS) proposed by Trichopoulou [42]. When the overall score was analyzed a reduction near 10% in the risk of diabetes was observed for those participants with the highest adherence to the MDS as compared to those with the minimum level of adherence. When the score was constructed eliminating olive oil, the attenuation of the effect was apparent with a reduction in the risk of only 3% and only marginally significant for those with the maximum adherence to this healthy dietary pattern. However, the effect was really found for Mediterranean Diet in which other important components beyond olive oil such as vegetables, fruits, or cereals could contribute to the reported effect. In fact, the association between Mediterranean dietary score and diabetes was attenuated when olive oil or other components such as alcohol or meat were excluded from the score.

Other important evidence comes from the PREDIMED trial as we have already confirmed. In a subgroup analysis, Salas-Salvadó et al. found that those participants receiving Mediterranean Diet plus extra-virgin olive oil showed a reduction near 40% in developing diabetes during a follow-up period of approximately 4 years [43].

Also in this trial, over a mean of 5.51 years of follow-up, participants with the highest intake of polyphenols had a 28% significant reduction in new-onset diabetes as compared to participants with the lowest total poly-phenol intake with a dose−response relationship. Moreover, some subclasses of polyphenols also were inversely associated with diabetes risk, including for total flavonoids (33% reduction in the risk), stilbenes (43% reduction in the risk), dihydroflavonols (41%), and flavanones (31%). Again, we have to highlight that the effect might not only be entirely due to extra-virgin olive oil but also to other sources of polyphenols such as fruits, vegetables, or legumes [44].

4.6.2 Hypertension

Some epidemiological studies have analyzed the role of different types of dietary fats and the incidence of hypertension or the changes in blood pres-sure failing to find any significant association. However, most of them have been conducted in the United States where, as we have mentioned previ-ously, the consumption is moderate and comes mainly from some types of meats and hence is correlated with the intake of SFA. Contrarily to the lack of association found for MUFA intake in several American studies such as the Nurses' Health Study, the Chicago Western Electric Company study, or the Multiple Risk Factor Intervention Trial, the few epidemiological studies conducted in Southern Europe show very different results suggesting a pro-tective role for MUFA or olive oil.

A cross-sectional analysis within the EPIC-Greece found that the MUFA/SFA intake ratio and olive oil consumption were inversely associated with systolic and diastolic blood pressure after taking into account other factors such as body mass index, energy intake, or physical activity of the partici-pants. However, the design of the cross-sectional studies hinders the estab-lishment of causal relationships [45].

It is remarkable the association found between olive oil consumption and hypertension within the SUN cohort study with prospective follow-up. In a sample of 5573 participants free of hypertension at baseline, the highest con-sumption of olive oil was associated to a 50% reduction in the risk of hyper-tension with a significant dose−response relationship but only among men. The authors concluded that probably the lack of association observed among women could be attributed to the overall lower incidence of hypertension found among females and the resulting lower statistical power [46].

Also the PREDIMED trial analyzed the role of extra-virgin olive oil in blood pressure. When the overall sample was analyzed, the groups assigned to both Mediterranean Diet plus extra-virgin olive oil or nuts showed a sig-nificant reduction in diastolic blood pressure during the follow-up period when were compared to the control group [47]. In a posterior analysis in a subsample of the trial ($n = 200$), both systolic and diastolic blood pressure decreased significantly after a 1-year dietary intervention with Mediterranean Diet supplemented with extra-virgin olive oil [48]. Thus, the reported reduction in cardiovascular disease that was more evident for stroke in other analysis within the PREDIMED trial among those participants assigned to Mediterranean Diet and extra-virgin olive oil, could be partly explained by the control of blood pressure exerted by this dietary pattern.

Other studies conducted in very controlled environments have suggested that olive oil could be used as a nonpharmacological approach in the treatment of hypertension. For example, Ferrara et al. found that a diet rich in extra-virgin olive oil was associated with a reduced need for antihypertensive medication compared with a diet enriched in sunflower oil. The authors concluded that a slight reduction in saturated fat intake, along with the use of extra-virgin olive oil, markedly lowers daily antihypertensive dosage requirement, possibly through enhanced nitric oxide levels stimulated by polyphenols [49]. Over again, the effect of olive oil in blood pressure would be not only mediated through its MUFA content.

Therefore, several epidemiological studies have concluded that polyphenols intake might decrease systolic and diastolic blood pressure in human. The mechanism could be related with the induction of vasodilatation via activation of the nitric oxide (NO) system (NO regulates blood pressure by dilating arteries). In this sense, Moreno-Luna et al. conducted a clinical trial in a sample of 24 women with high–normal blood pressure or stage 1 essential hypertension to assess the role of polyphenol rich olive oil as compared to refined oil (polyphenol free olive oil) on blood pressure. After 2 months of intervention, a significant reduction in blood pressure (both systolic and diastolic pressure) was found for the group assigned to polyphenol rich olive oil. Moreover, after the polyphenol rich olive oil diet a decrease in serum oxidized LDL and plasma C-reactive protein was observed as well as an increase in plasma nitrates and nitrites (metabolites of NO). The state of hypertension is often associated with increased vascular oxidative stress and inflammation. So, these findings could explain the olive oil polyphenols effects on blood pressure through their antiinflammatory and antioxidant properties, improvement of endothelial function and thereby a restoring of vascular reactivity [50].

Likewise, also the PREDIMED trial has analyzed the role of polyphenols in blood pressure. In a sample of 200 participants from the trial, systolic and diastolic blood pressure decreased significantly after a 1-year dietary intervention with Mediterranean Diet supplemented with extra-virgin olive oil as we have already mentioned. These changes in blood pressure were associated with a significant increase in the level of polyphenols excreted in urine and plasma NO. The statistically significant increases in plasma NO were associated with a reduction in systolic and diastolic blood pressure levels, adding to the growing evidence that polyphenols might protect the cardiovascular system by improving the endothelial function and enhancing endothelial synthesis of NO [48].

4.6.3 Obesity

Although olive oil is an energy-dense food with 100% of fat content, both observational and dietary intervention trials consistently have shown that Mediterranean Diet rich in olive oil does not contribute to obesity. However, only a few epidemiological studies have analyzed the role of olive oil independently of the Mediterranean Diet.

The SUN cohort study found that olive oil consumption was associated neither to overweight or obesity nor with weight gain in a sample of 7368 participants followed up for a mean of 28.5 months [51].

In an analysis within a subsample of the PREDIMED trial and published in 2009, Razquin et al. found that the adherence to the Mediterranean Diet supplemented with extra-virgin olive oil was associated with better plasmatic antioxidant profile and that this antioxidant profile was inversely related with weight gain after 3 years of follow-up [52]. However, in a recent analysis within the PREDIMED trial, Alvarez et al. did not find any significant difference among intervention groups in several anthropometric and body fat measures after one year of intervention in a subsample composed of 305 participants [53]. When the overall sample was analyzed, a higher reduction in body weight after 5 years of follow-up was observed for those participants assigned to the intervention group with extra-virgin olive oil compared to the control group. The authors concluded that the findings lend support to advice not restricting intake of healthy fats such as extra-virgin olive oil for body weight maintenance [54].

This lack of weight gain and the improvement in overweight or obesity exerted by olive oil may be due to the increased fat oxidation after ingestion, diet-induced thermogenesis and overall daily energy expenditure induced by olive oil. Other authors have hypothesized that olive oil could exert positive effects on fat redistribution. A short-term experimental study involving individuals with obesity and insulin resistance showed that a Mediterranean Diet rich in extra-virgin olive oil prevented accumulation of central body fat compared with a low-fat diet without affecting body weight [55].

4.6.4 Metabolic Syndrome

The metabolic syndrome is a multicomponent disorder characterized by hypertriglyceridemia, hyperglycemia, low HDL cholesterol, hypertension, and abdominal obesity. The association between olive oil consumption and different components of metabolic syndrome has been described along this chapter. Nevertheless, only a few cross-sectional, cohort, and intervention studies have analyzed the relationship between Mediterranean Diet adherence (with olive oil as the main source of fat) and the prevalence or incidence of metabolic syndrome [56] and none of them have specifically analyzed the role of olive oil. In this sense, the role of a Mediterranean Diet supplemented with extra-virgin olive oil in metabolic syndrome was analyzed in the PREDIMED trial. The adherence to this pattern was associated with a significant reversion of metabolic syndrome compared to the advice on following a low-fat diet (the probability of reversion was 35% higher among those assigned to the Mediterranean Diet plus extra-virgin olive oil than among those participants in the control diet). However, the researchers found no beneficial effect of this pattern on incidence of new-onset metabolic syndrome [57]. Also in other subsample of participants from the PREDIMED trial, Mayneris-Perxachs et al. examined the association between plasma levels of several fatty acids including oleic acid (biomarker of the consumption of olive oil in this sample) and the incidence, reversion, and changes in prevalence of metabolic syndrome after one year of intervention. Using the biomarkers of foods supplied to the intervention groups they found that the incidence and reversion rates of metabolic syndrome progressed inversely and in parallel, respectively, to increases in oleic acid and α-linolenic acid (biomarker of nuts consumption), which resulted in significant differences in the changes in prevalence of metabolic syndrome across the quartiles of changes in both plasma fatty acids [58].

4.7 HOW MUCH OIL IS RECOMMENDED A DAY?

One of the main characteristics of the Mediterranean Diet is its high fat intake, predominantly in the form of olive oil. Olive oil makes the diet more palatable and easier to follow than a low-fat diet, in which fat content is significantly restricted, including olive oil. Based on the scientific evidence from the PREDIMED study, in which the participants consumed an average of 50 ml of extra-virgin olive oil per day, the daily recommendation to follow a healthy Mediterranean Diet is between 4 and 6 tablespoons. This amount includes both raw (salad dressing, vegetables, toast, etc.) and cooked (stews, fried food, etc.) forms of consuming olive oil.

4.8 CONSUMER TIPS: HOW TO TAKE ADVANTAGE OF OLIVE OIL

At first, it may seem like extra-virgin olive oil is much more expensive than other types of oil, but ultimately it can prove to be more economical and more beneficial in the long run. This is explained by the need to use a lesser quantity of oil, due to its expanding properties and stronger flavor.

Olive oil is more resistant to high temperatures than other oils for cooking, making it more suitable oil for frying. If the olive oil that was used for frying is not burned, it can be reused several times. Other vegetable oils, on the contrary, burn quicker and their reuse is not recommended.

In addition, olive oil is absorbed less by food than other seed oils. This results in dishes with a lower fat content and lower caloric intake than those that use other types of cooking oil.

To prevent the alteration of the health properties, it is highly advised to refrain from high heat cooking methods and to store food in opaque or colored containers, to avoid direct contact with the light. It is not recommended to mix fresh olive oil with used olive oil or other kinds of oils. One last recommendation for oil is to avoid frying food with high water content, because water accelerates the oil's rate of degradation.

4.9   HOW CAN I INCORPORATE OLIVE OIL IN MY DIET?

- Incorporate the use of olive oil at home, not only when using raw (as salad dressing), but also for cooking and frying.
- Start your day off with a healthy breakfast of toast drizzled with olive oil.
- Use olive oil for salad dressings instead of unhealthier dressings.
- Try using it at the dinner table instead of butter.
- Replace olive oil for other fats such as butter, margarine, or cream when making sauces.
- You can enhance the flavor of your dishes with olive oil previously soaked with herbs.
- Avoid industrial baked goods and sources of animal fats and cholesterol such as bacon. Eat comfortably seated and in moderation. Use olive oil to bake homemade pastries, cakes and muffins.

REFERENCES

[1]   Kris-Etherton PM. AHA Science Advisory. Monounsaturated fatty acids and risk of cardio-vascular disease. American Heart Association. Nutrition Committee. Circulation, 100. 1990. p. 1253−8.
[2]   Gardner CD, Kraemer HC. Monounsaturated versus polyunsaturated dietary fat and serum lipids: a meta-analysis. Arterioscler Thromb Vasc Biol 1995;15:1917−27.
[3]   Gómez M, Vila J, Elosua R, Molina L, Bruguera J, Sala J, et al. Relationship of lipid oxi-dation with subclinical atherosclerosis and 10-year coronary events in general population. Atherosclerosis 2014;232:134−40.
[4]   Covas MI, Nyyssönen K, Poulsen HE, Kaikkonen J, Zunft HJ, Kiesewetter H, et al. The effect of polyphenols in olive oil on heart disease risk factors: a randomized trial. Ann Intern Med 2006;145:333−41.
[5]   Covas MI, Konstantinidou V, Fito M. Olive oil and cardiovascular health. J Cardiovasc Pharmacol 2009;54:477−82.
[6]   Fitó M, Guxens M, Corella D, Sáez G, Estruch R, de la Torre R, et al. PREDIMED Study Investigators. Effect of a traditional Mediterranean diet on lipoprotein oxidation: a randomized controlled trial. Arch Intern Med 2007;157:1195−203.

[7] Fitó M, Estruch R, Salas-Salvadó J, Martínez-Gonzalez MA, Arós F, Vila J, et al. Effect of the Mediterranean diet on heart failure biomarkers: a randomized sample from the PREDIMED trial. Eur J Heart Fail 2014;16:543–50.

[8] Quiles JL, Ochoa JJ, Ramirez-Tortosa C, Battino M, Huertas JR, Martín Y, et al. Dietary fat type (virgin olive vs. sunflower oils) affects age-related changes in DNA double-strand-breaks, antioxidant capacity and blood lipids in rats. Exp Gerontol 2004;39:1189–98.

[9] Estruch R, Martínez-González MA, Corella D, Salas-Salvadó J, Ruiz-Gutiérrez V, Covas MI, et al. Effects of a Mediterranean-style diet on cardiovascular risk factors: a randomized trial. Ann Intern Med 2006;145:1–11.

[10] Medina-Remón A, Casas R, Tressserra-Rimbau A, Ros E, Martínez-González MA, Fitó M, et al. Polyphenol intake from a Mediterranean diet decreases inflammatory biomarkers related to atherosclerosis: A sub-study of The PREDIMED trial. Br J Clin Pharmacol 2017;83(1):114–28.

[11] Schwingshackl L, Christoph M, Hoffmann G. Effects of olive oil on markers of inflammation and endothelial function—a systematic review and meta-analysis. Nutrients 2015;7:7651–75.

[12] Pacheco YM, Bermúdez B, López S, Abia R, Villar J, Muriana FJ. Minor compounds of olive oil have postprandial anti-inflammatory effects. Br J Nutr 2007;98:260–3.

[13] Fitó M, Cladellas M, de la Torre R, Martí J, Muñoz D, Schröder H, et al. Anti-inflammatory effect of virgin olive oil in stable coronary disease patients: a randomized, crossover, controlled trial. Eur J Clin Nutr 2008;62:570–4.

[14] Mena MP, Sacanella E, Vazquez-Agell M, Morales M, Fitó M, Escoda R, et al. Inhibition of circulating immune cell activation: a molecular antiinflammatory effect of the Mediterranean diet. Am J Clin Nutr 2009;89:248–56.

[15] Delgado-Lista J, Garcia-Rios A, Perez-Martinez P, Lopez-Miranda J, Perez-Jimenez F. Olive oil and haemostasis: platelet function, thrombogenesis and fibrinolysis. Curr Pharm Design 2011;17:778–85.

[16] Keys A, Menotti A, Karvonen MJ, Aravanis C, Blackburn H, Buzina R, et al. The diet and 15-year death rate in the seven countries study. Am J Epidemiol 1986;124:903–15.

[17] Tunstall-Pedoe H, Kuulasmaa K, Amouyel P, Arveiler D, Rajakangas AM, Pajak A. Myocardial infarction and coronary deaths in the World Health Organization MONICA Project. Registration procedures, event rates, and case-fatality rates in 38 populations from 21 countries in four continents. Circulation 1994;90:583–612.

[18] Fernández-Jarne E, Martínez-Losa E, Prado-Santamaría M, Brugarolas-Brufau C, Serrano-Martínez M, Martínez-González MA. Risk of first non-fatal myocardial infarction negatively associated with olive oil consumption: a case-control study in Spain. Int J Epidemiol 2002;31:474–80.

[19] Kontogianni MD, Panagiotakos DB, Chrysohoou C, Pitsavos C, Zampelas A, Stefanidis C. The impact of olive oil consumption pattern on the risk of acute coronary syndromes: the CARDIO2000 case-control study. Clin Cardiol 2007;30:125–9.

[20] Bertuzzi M, Tavani A, Negri E, La Vecchia C. Olive oil consumption and risk of non-fatal myocardial infarction in Italy. Int J Epidemiol 2002;31:1274–7.

[21] Gramenzi A, Gentile A, Fasoli M, Negri E, Parazzini F, La Vecchia C. Association between certain foods and risk of acute myocardial infarction in women. BMJ 1990;300:771–3.

[22] Buckland G, Mayén AL, Agudo A, Travier N, Navarro C, Huerta JM, et al. Olive oil intake and mortality within the Spanish population (EPIC-Spain). Am J Clin Nutr 2012;96:142–9.

[23] Buckland G, Travier N, Barricarte A, Ardanaz E, Moreno-Iribas C, Sánchez MJ, et al. Olive oil intake and CHD in the European Prospective Investigation into Cancer and Nutrition Spanish cohort. Br J Nutr 2012;108:2075−82.

[24] Dilis V, Katsoulis M, Lagiou P, Trichopoulos D, Naska A, Trichopoulou A. Mediterranean diet and CHD: the Greek European Prospective Investigation into Cancer and Nutrition cohort. Br J Nutr 2012;108:699−709.

[25] Misirli G, Benetou V, Lagiou P, Bamia C, Trichopoulos D, Trichopoulou A. Relation of the traditional Mediterranean diet to cerebrovascular disease in a Mediterranean population. Am J Epidemiol 2012;176:1185−92.

[26] Samieri C, Féart C, Proust-Lima C, Peuchant E, Tzourio C, Stapf C, et al. Olive oil consumption, plasma oleic acid, and stroke incidence: the Three-City Study. Neurology 2011;77:418−25.

[27] Bendinelli B, Masala G, Saieva C, Salvini S, Calonico C, Sacerdote C, et al. Fruit, vegetables, and olive oil and risk of coronary heart disease in Italian women: the EPICOR Study. Am J Clin Nutr 2011;93:275−83.

[28] Guasch-Ferré M, Hu FB, Martínez-González MA, Fitó M, Bulló M, Estruch R, et al. Olive oil intake and risk of cardiovascular disease and mortality in the PREDIMED Study. BMC Med 2014;12:78.

[29] Barzi F, Woodward M, Marfisi RM, Tavazzi L, Valagussa F, Marchioli R. Mediterranean diet and all-causes mortality after myocardial infarction: results from the GISSI-Prevenzione trial. Eur J Clin Nutr 2003;57:604−11.

[30] de Lorgeril M, Salen P, Martin JL, Monjaud I, Delaye J, Mamelle N. Mediterranean diet, traditional risk factors, and the rate of cardiovascular complications after myocardial infarction: final report of the Lyon Diet Heart Study. Circulation 1999;99:779−85.

[31] Estruch R, Ros E, Salas-Salvadó J, Covas MI, Corella D, Arós F, et al. Primary prevention of cardiovascular disease with a Mediterranean diet. N Engl J Med 2013;368:1279−90.

[32] Ruiz-Canela M, Estruch R, Corella D, Salas-Salvadó J, Martínez-González MA. Association of Mediterranean diet with peripheral artery disease: the PREDIMED randomized trial. JAMA 2014;311:415−17.

[33] Martínez-González MÁ, Toledo E, Arós F, Fiol M, Corella D, Salas-Salvadó J, et al. Extravirgin olive oil consumption reduces risk of atrial fibrillation: the PREDIMED (Prevención con Dieta Mediterránea) trial. Circulation 2014;130:18−26.

[34] Martínez-González MA, Dominguez LJ, Delgado-Rodríguez M. Olive oil consumption and risk of CHD and/or stroke: a meta-analysis of case-control, cohort and intervention studies. Br J Nutr 2014;112:248−59.

[35] Schwingshackl L, Hoffmann G. Monounsaturated fatty acids, olive oil and health status: a systematic review and meta-analysis of cohort studies. Lipids Health Dis 2014;13:15.

[36] Chowdhury R, Warnakula S, Kunutsor S, Crowe F, Ward HA, Johnson L, et al. Association of dietary, circulating, and supplement fatty acids with coronary risk: a systematic review and meta-analysis. Ann Intern Med 2014;160:398−406.

[37] Jakobsen MU, O'Reilly EJ, Heitmann BL, Pereira MA, Bälter K, Fraser GE, et al. Major types of dietary fat and risk of coronary heart disease: a pooled analysis of 11 cohort studies. Am J Clin Nutr 2009;89:1425−32.

[38] Tresserra-Rimbau A, Rimm EB, Medina-Remón A, Martínez-González MA, de la Torre R, Corella D, et al. Inverse association between habitual polyphenol intake and incidence of cardiovascular events in the PREDIMED study. Nutr Metab Cardiovasc Dis 2014;24:639−47.

[39] Soriguer F, Rojo-Martínez G, Goday A, Bosch-Comas A, Bordiú E, Caballero-Díaz F, et al. Olive oil has a beneficial effect on impaired glucose regulation and other cardiometabolic risk factors. Diabetes study. Eur J Clin Nutr 2013;67:911–16.

[40] Guasch-Ferré M, Hruby A, Salas-Salvadó J, Martínez-González MA, Sun Q, Willett WC, et al. Olive oil consumption and risk of type 2 diabetes in US women. Am J Clin Nutr 2015;102:479–86.

[41] Marí-Sanchis A, Beunza JJ, Bes-Rastrollo M, Toledo E, Basterra- Gortariz FJ, Serrano-Martínez M, et al. Olive oil consumption and incidence of diabetes mellitus, in the Spanish sun cohort. Nutr Hosp 2011;26:137–43.

[42] InterAct Consortium, Romaguera D, Guevara M, Norat T, Langenberg C, Forouhi NG, et al. Mediterranean diet and type 2 diabetes risk in the European Prospective Investigation into Cancer and Nutrition (EPIC) study: the InterAct project. Diabetes Care 2011;34:1913–18.

[43] Salas-Salvadó J, Bulló M, Estruch R, Ros E, Covas MI, Ibarrola-Jurado N, et al. Prevention of diabetes with Mediterranean diets: a subgroup analysis of a randomized trial. Ann Intern Med 2014;160:1–10.

[44] Tresserra-Rimbau A, Guasch-Ferré M, Salas-Salvadó J, Toledo E, Corella D, Castañer O, et al. Intake of total polyphenols and some classes of polyphenols is inversely associated with diabetes in elderly people at high cardiovascular disease risk from the PREDIMED trial. J Nutr 2017; [Epub ahead of print].

[45] Psaltopoulou T, Naska A, Orfanos P, Trichopoulos D, Mountokalakis T, Trichopoulou A. Olive oil, the Mediterranean diet, and arterial blood pressure: the Greek European Prospective Investigation into Cancer and Nutrition (EPIC) study. Am J Clin Nutr 2004;80:1012–18.

[46] Alonso A, Martínez-González MA. Olive oil consumption and reduced incidence of hypertension: the SUN study. Lipids 2004;39:1233–8.

[47] Toledo E, Hu FB, Estruch R, Buil-Cosiales P, Corella D, Salas-Salvadó J, et al. Effect of the Mediterranean diet on blood pressure in the PREDIMED trial: results from a randomized controlled trial. BMC Med 2013;1:207.

[48] Medina-Remón A, Tresserra-Rimbau A, Pons A, Tur JA, Martorell M, Ros E, et al. Effects of total dietary polyphenols on plasma nitric oxide and blood pressure in a high cardiovascular risk cohort. The PREDIMED randomized trial. Nutr Metab Cardiovasc Dis 2015;25:60–7.

[49] Ferrara LA, Raimondi AS, d'Episcopo L, Guida L, Dello-Russo A, Marotta T. Olive oil and reduced need for antihypertensive medications. Arch Intern Med 2000;160:837–42.

[50] Moreno-Luna R, Muñoz-Hernandez R, Miranda ML, Costa AF, Jimenez-Jimenez L, Vallejo-Vaz AJ, et al. Olive oil polyphenols decrease blood pressure and improve endothelial function in young women with mild hypertension. Am J Hypertens 2012;25:1299–304.

[51] Bes-Rastrollo M, Sánchez-Villegas A, de la Fuente C, de Irala J, Martinez JA, Martínez-González MA. Olive oil consumption and weight change: the SUN prospective cohort study. Lipids 2006;41:249–56.

[52] Razquin C, Martinez JA, Martinez-Gonzalez MA, Mitjavila MT, Estruch R, Marti A. A 3 years follow-up of a Mediterranean diet rich in virgin olive oil is associated with high plasma antioxidant capacity and reduced body weight gain. Eur J Clin Nutr 2009;63:1387–93.

[53] Álvarez-Pérez J, Sánchez-Villegas A, Díaz-Benítez EM, Ruano-Rodríguez C, Corella D, Martínez-González MÁ, et al. Influence of a Mediterranean dietary pattern on body fat

distribution: results of the PREDIMED-Canarias Intervention Randomized Trial. J Am Coll Nutr 2016;35(6):568−80.

[54] Estruch R, Martínez-González MA, Corella D, Salas-Salvadó J, Fitó M, Chiva-Blanch G, et al. Effect of a high-fat Mediterranean Diet on bodyweight and waist circumference: a prespecified secondary outcomes analysis of the PREDIMED randomised controlled trial. Lancet Diabetes Endocrinol 2016;4(8):666−76.

[55] Paniagua JA, Gallego de la Sacristana A, Romero I, Vidal-Puig A, Latre JM, Sanchez E, et al. Monounsaturated fat-rich diet prevents central body fat distribution and decreases postprandial adiponectin expression induced by a carbohydrate-rich diet in insulin-resistant subjects. Diabetes Care 2007;30:1717−23.

[56] Esposito K, Giugliano D. Mediterranean diet and the metabolic syndrome: the end of the beginning. Metab Syndr Relat Disord 2010;8:197−200.

[57] Babio N, Toledo E, Estruch R, Ros E, Martínez-González MA, Castañer O, et al. Mediterranean diets and metabolic syndrome status in the PREDIMED randomized trial. CMAJ 2014;186:E649−57.

[58] Mayneris-Perxachs J, Sala-Vila A, Chisaguano M, Castellote AI, Estruch R, Covas MI, et al. Effects of 1-year intervention with a Mediterranean diet on plasma fatty acid composition and metabolic syndrome in a population at high cardiovascular risk. PLoS One 2014;9:e85202.

Chapter 5

A Small Handful of Mixed Nuts

Maira Bes-Rastrollo and Ana Sánchez-Tainta

5.1 INTRODUCTION

Walnuts, hazelnuts, pistachios, almonds, cashews, etc. were already consumed in the Roman Empire and they are still staple foods in many recipes around the world.

However, their high fat content (between 50% and 70% of their weight) confers them a bad reputation in terms of weight maintenance, and as a consequence, general population has avoided them.

Overall, nuts are a good source of healthy cardiovascular fat omega-3 and omega-6, essential fatty acids (the human body does not produce them); in addition, they have no cholesterol. Specially, walnuts are rich in polyunsaturated fatty acids omega-3, with healthy demonstrated benefits; hazelnuts and almonds stand out for their content in monounsaturated fatty acids, like the kind of fatty acids present in olive oil.

Nuts consumption, when they are consumed instead of other foods with the same quantity of saturated fatty acids, reduce plasma LDL cholesterol ("bad" cholesterol), prevent diabetes, and improve blood pressure and symptoms of inflammatory diseases.

The reason of these findings is based on the high nutrient content of nuts. They are one of the most nutritious foods. They provide proteins from vegetal origin, they are a good source of vitamins including folic acid, niacin, vitamin B6, vitamin E and A, and minerals such as magnesium, copper, zinc, selenium, phosphorous, calcium, and potassium. Some of these components (such as vitamin E and selenium) have a high antioxidant effect, similar to others phytoquemical compounds found in nuts. Fiber is also important in the composition of nuts, and confers healthy benefits. An adequate consumption of fiber helps to improve and prevent constipation, regulate blood glucose and cholesterol levels, and maintain healthy weight and prevent obesity.

Nuts are a source of L-arginine, an amino acid substrate for endothelium-derived NO synthesis, with the consequence of dilating blood vessels, and therefore improving arterial blood circulation.

The Prevention of Cardiovascular Disease through The Mediterranean Diet.
DOI: http://dx.doi.org/10.1016/B978-0-12-811259-5.00005-6

It is important to note that most of the micronutrients are located in the pellicle or outer soft shells, which means that it is better to consume them raw with outer soft shells than consume them peeled or roasted.

Because of their excellent nutritional profile, nuts are foods very recommended for everybody: children and adolescents, pregnant and breastfeeding women, and sportsman and sportswomen. For elderly people are a good source of nutrients that helps them to avoid deficiencies.

This chapter summarizes current knowledge on the topic of nuts as a healthy food in the context of a Mediterranean Diet.

5.2   SCIENTIFIC EVIDENCE

Tree nuts are defined as dry fruits with one seed in which the ovary wall becomes hard at maturity. There are several kinds of tree nuts including those cited at the beginning and also Brazil nuts, pecans, and pine nuts. From a botanical point of view, peanuts are legumes, but they have a similar nutrient profile to tree nuts, for this reason, they usually are included in the context of nuts from a consumer perspective [1].

The Adventist Health Study cohort was the first study to suggest a protective effect of nut consumption on coronary heart disease (CHD) in 1992 [2]. These findings were replicated in a landmark randomized clinical trial (RCT) conducted by Sabate et al. They found an inverse association between walnut intake and serum cholesterol levels [3].

The proposal of nuts as a healthy food is supported by both epidemiological observations showing that frequency of nut consumption was inversely associated with incident CHD and diabetes and from several randomized clinical trials demonstrating beneficial effects of nuts consumption on blood lipids and other intermediators of CHD [4].

The Mediterranean Diet is defined as the dietary pattern of those people living in the olive tree-growing areas of the Mediterranean basin before mid-1960s [5]. Nut consumption is an important component of this pattern. Its consumption jointly with olive oil consumption as the main source of fat make the Mediterranean diet a high-fat dietary pattern (30% and 45% of the total energy intake). This dietary pattern is also characterized by a high consumption of fruits, vegetables, legumes, and whole-grain cereals; and with low consumption of meat and meat products; and a relatively low consumption of dairy products, usually in the form of long-preserved cheeses. Alcohol consumption in moderation in the form of wine during meals is also a characteristic of this pattern. Finally, moderate consumption of fish, depending on the distance from the sea, in the past, is another characteristic of this pattern [6].

5.2.1   The Secret of Nuts: Their Nutrient Content

Nuts are one of the plant-based food richest in fat after vegetable oils. Nuts have a high total fat content, ranging from 68% for pine nuts to 46% in pistachios (Table 5.1). The good point is that the fatty acid composition of the

TABLE 5.1 Nutrients Composition of Nuts in 100 g of Edible Portion

	Calories (Kcal)	Proteins (g)	SFA (g)	MUFA (g)	PUFA (g)	CH (g)	Fiber (g)	Se (mcg)	P (mg)	Mg (mg)	Na (mg)	K (mg)	Vit A (IU)	Vit E (mg)	Vit B6 (mg)
Almonds	575	21	4	31	12	20	12	2.5	484	268	1	705	1	26	0.14
Brazil nuts	656	14	15	25	21	12	8	1917	725	1223	3	659	0	6	0.10
Cashews	553	18	8	24	8	30	3	20	593	292	12	660	0	0.9	0.42
Hazelnuts	628	15	5	46	8	17	10	2.4	290	163	0	680	20	15	0.56
Macadamias	718	8	12	59	1.5	14	9	3.6	188	130	5	368	0	0.54	0.28
Pecans	691	9	6	41	22	14	10	3.8	277	121	0	410	56	1.4	0.21
Pine nuts	673	14	5	19	34	13	4	0.7	575	251	2	597	29	9.33	0.09
Pistachios	557	21	5.4	23	14	28	10	7	490	121	1	1025	553	2.3	1.7
Walnuts	654	15	6	9	47	14	7	4.90	346	158	2	441	20	0.7	0.54

SFA, Saturated fatty acids; *MUFA*, monounsaturated fatty acids; *PUFA*, polyunsaturated fatty acids; *CH*, carbohydrates.
Source: Adapted from International and Dried Fruit Foundation.

TABLE 5.2 Mechanisms of Cardiometabolic Protection Exerted by Nutrients of Nuts

Decrease cholesterol
- Unsaturated fatty acids
- Fiber
- Phytoesterols

Decrease oxidation and inflammation
- Phytoesterols
- Tocopherols
- Polyphenols

Increase NO → Decrease blood pressure
- Polyphenols
- Vegetable protein (L-arginine)
- Minerals (magnesium, potassium, calcium)

Increase glucose control
- Unsaturated fatty acids
- Fiber
- Minerals (magnesium, potassium, calcium)
- Vegetable protein (L-arginine)

fat from nuts is beneficial from a health perspective because their saturated content is low (4%−6%), and half of the total fat content is unsaturated fat, monounsaturated fatty acids (MUFAs) in most nuts, and the polyunsaturated fatty acid linoleic acid and α-linolenic acid, the plant omega-3 fatty acids, in walnuts [1].

Second, with the exception of chestnuts, nuts contain few carbohydrates and, thus, contribute little to postprandial glycemia [7].

Third, frequent nut consumption has a positive association with plasma adiponectin and an inverse association with circulating inflammatory cytokines [8]. Fourth, other constituents of nuts such as plant-origin proteins, L-arginine, folate, fiber, and phytoesterols contribute to their salutary effects [9] (Table 5.2). Nuts are also rich in minerals (e.g., calcium, magnesium, and potassium) [9], which are associated with decreased overall risk of mortality and cardiovascular disease (CVD) [10].

Recently, a cross-sectional study of the Nurses' Health Study concluded that frequent nut consumption was associated with lower amounts of C-reactive protein and IL-6 [11].

5.2.2 Epidemiological Evidence on Healthy Effects of Nut Consumption: Observational Studies

Luo et al. [10] in 2014 published a systematic review and meta-analysis of prospective cohort studies on nut consumption and risk of type 2 diabetes,

CVD, and all-cause mortality. They found 18 prospective cohort studies with 31 reports, 5 on diabetes, 6 on CHD, 5 on stroke, 4 on total CVD, and 11 on all-cause mortality.

Pooled results from this systematic review including four reports with age-adjusted estimates suggested an inverse association between nut intake and diabetes [relative risk (RR) = 0.88; 95% confidence interval (CI): 0.84−0.92] for an increment of 1 serving per day. However, when they included only those reports adjusted for body mass index, the inverse association disappeared (RR = 1.03; 95% CI: 0.91−1.16), indicating that the association was largely mediated through body mass index.

For overall CVD an increment of 1 serving per day of nuts reduced the risk of CVD 29% (RR = 0.71; 95% CI: 0.59−0.85), when they removed those reports that included peanut butter the reduction was slightly higher (36%; RR = 0.64; 95% CI: 0.43−0.97). They found similar results for only CHD. The results assessing only stroke for the comparison of extreme quantiles showed a lower nonstatistically significant risk reduction (RR = 0.91; 95% CI: 0.81−1.02). However, in stratified analyses by sex, women presented a statistically significant 13% reduction (RR = 0.87; 95% CI: 0.77−0.98), whereas among men the association was less apparent and not statistically significant (RR = 0.95; 95% CI: 0.82−1.11).

For all-cause mortality, the increment of 1 serving/d in nut intake was associated with lower risk of death (RR = 0.83; 95% CI: 0.76−0.91); however, with a moderately high heterogeneity ($I^2 = 62.1\%$, $P = 0.032$).

In summary, this systematic review concluded that nut intake reduced the risk of overall CVD, CHD, stroke in women, and all-cause mortality, but not with diabetes and total stroke.

These findings are fairly consistent with those in another systematic review and meta-analysis that included 25 observational studies, including two case-control studies and two reports from the PREDIMED (PREvención con DIeta MEDiterránea in Spanish) trial. They found that eating nuts was associated with reduced risks of diabetes, and CHD, but not of stroke [12].

In a recent meta-analysis published at the end of 2016 in BMC Medicine, Aune et al. [13] included twenty studies (with 29 publications). They concluded that an increase of 28 grams/d in nut intake was associated with reduced risk of CVD (RR = 0.79; 95% CI: 0.70−0.88; 12 publications), including CHD (RR = 0.71; 95% CI: 0.63−0.80; 11 publications), but not stroke (RR = 0.93; 95% CI: 0.83−1.05; 11 publications), total cancer (RR = 0.85; 95% CI: 0.76−0.94; 8 publications), all-cause mortality (RR = 0.78; 95% CI: 0.72−0.84; 15 publications), and mortality from respiratory disease (RR = 0.48; 95% CI: 0.26−0.89; 3 publications), mortality from diabetes (RR = 0.61; 95% CI: 0.43−0.88; 4 publications), and mortality from infections (RR = 0.25; 95% CI: 0.07−0.85), although for this last outcome there were only available two publications [13].

Since always is possible the existence of residual confounding in observational results, and those who consume more nuts usually have healthier lifestyles, results from the Golestan Cohort Study based in north-eastern Iran provide evidence for a 29% reduction in total mortality [hazard ratio (HR): 0.71; 95% CI: 0.58−0.86; P for trend < 0.001] for those participants in the highest category of nuts consumption (> = 3 servings/week) in comparison of those who did not consume nuts, even in a population whose nut consumption does not track with healthier lifestyles [14].

Findings from two cross-sectional analyses based the first one on the National Health and Nutrition Examination Survey (NHANES) [15] and the other in the PREDIMED trial [16], and results from the Spanish SUN (Seguimiento Universidad de Navarra in Spanish) cohort [17] suggested also a protective role of nuts on metabolic syndrome [18].

5.2.3 Epidemiological Evidence on Healthy Effects of Nut Consumption: Clinical Trials

Most RCTs assessing the healthy effect of nut consumption have been short and based on intermediate factors that help to understand the underlying mechanisms for cardioprotective effect of nuts in observational studies.

The effect of nut consumption on blood lipid levels was assessed in a pooled analyses of 25 intervention trials showing a dose-response cholesterol lowering effect [19]. Previously, similar results were found in a meta-analysis of 13 clinical trials with diet supplemented with walnuts showing a decrease in total and LDL cholesterol [20].

Because of these beneficial effects, the protective effect of frequent nut consumption on overall CVD, CHD, and all-cause mortality is supported by biological plausibility.

Long-term, double-blind, RCTs provide the best evidence on the effect of food and dietary patterns and health outcomes. But not always it is possible to have available scientific evidence from these epidemiological studies. We are very fortunate to have the available results from the PREDIMED trial, a Spanish multicentered primary prevention cardiovascular trial of 7447 high cardiovascular risk participants. Approximately one third of participants were advised to follow a Mediterranean Diet supplemented with nuts consumption (15 g walnuts, 7.5 g almonds, and 7.5 g hazelnuts). The results showed that those participants allocated to the Mediterranean Diet rich in nuts after a median follow-up of 4.8 years reduced 30% their risk of stroke, total CVD, and cardiovascular mortality compared with the control group (HR = 0.70; 95% CI: 0.53−0.94). Of note, the nut supplemented Mediterranean Diet reduced 49% the risk of stroke (HR = 0.61; 95% CI: 0.44−0.86) [21]. This intervention also reduced systolic blood pressure [22], risk of metabolic syndrome [23], and in the case of diabetes results showed a nonsignificant 19% reduction [24]. Both Mediterranean dietary interventions (rich in olive oil and rich in nuts) of the PREDIMED showed an inverse association with another cardiovascular outcome: peripheral arterial disease [25]. Moreover, in a sub-sample of 447 participants of the PREDIMED, those allocated to Mediterranean Diet supplemented with nuts increased their memory cognitive function during follow-up whereas those in the control group decreased this function, being this difference statistically significant [26].

5.2.4   Eating Nuts Does Not Make You Fat!

Thanks to the available scientific evidence, today it is possible to support that eating nuts does not make you fat, despite their high fat content [27,28]. In the PREDIMED trial, those more than 2000 participants who consumed nuts daily during a mean of 5 years, on average did not increase significantly their weight. Moreover, results showed beneficial effects for waist circumference and body mass index related to nuts consumption [21].

There are two main reasons to explain these results:

- Satiety power: their fat and fiber content make us full and satisfy when we consume them. This avoids the subsequent snacking with less healthy foods between hours. They account for up to 75% of the energy they provide [29].
- Fat malabsorption: fat in nuts are in cellular membranes that are incompletely digested even after mastication. Not all fat content is absorbed. It has been documented as fecal fat excretion [30].

Results from prospective cohort studies showed that frequent nut consumption was associated with lower weight gain or obesity using data of the SUN prospective cohort study and the Nurses' Health Study II. After adjusting for potential confounders, those who consumed nuts two or more times per week had a 29% reduced risk of weight gain ($> = 3$ kg per year) (adjusted odds ratio = 0.71; 95% CI: 0.54−0.93) [31] or a 38% lower risk of obesity (adjusted HR = 0.62; 95% CI: 0.32−0.99) [32] in comparison with those who never or almost never consumed nuts.

In a meta-analysis of RCTs of diets rich in nuts versus control diet, showed no evidence of increased adiposity with frequent nut consumption [33].

In the recent published report on long-term body weight change of the PREDIMED trial, despite dietary fat consumption was higher in the Mediterranean Diet groups than in the control diet (42% vs. 37% total energy intake), they did not find greater weight gain, and in agreement with the previous observational results they found some evidence for less weight gain and lower waist circumference. Those participants in the Mediterranean Diet rich in nuts they presented a mean weight difference after a median of 4.8 years of follow-up of $-0 \cdot 08$ kg ($-0 \cdot 50$ to $0 \cdot 35$; $P = 0 \cdot 730$) and $-0 \cdot 55$ cm ($-1 \cdot 16$ to $-0 \cdot 06$; $P = 0 \cdot 048$) of waist circumference compared to the control group [34].

5.3 RECOMMENDATIONS

The recommendation would be to consume daily nuts, or at least three days per week. It is possible to consume any variety; however, it is better to consume mixed nuts in order to take advantage of different nutrients from each one. The recommended quantity is 20–30 g per day, equivalent to a handful of nuts without nutshell. The key point is to control the quantity of nuts consumed in order to consume them in a healthy manner without affecting weight gain.

To avoid the temptation to consume too much nuts, we recommend:

- Do not eat directly from the bag. Take only the needed quantity and put away the rest in another place beyond our reach, especially if we are going to consume them watching television or in the cinema.
- To mix them with other food products (yogurt, salads, pasta, etc.). Using this way we will only eat those included in the dish.
- To carry them as snacks.
- To buy them with nutshell. As a consequence of cracking them, we will eat them slowly and probably we will eat less.

5.3.1 Consumption Advice

To take the maximum advantage of their healthy properties, the following are recommended:

- Eat them in their natural form, not toasted or fried, because heat affects their healthy components, although a recent RCT with 72 participants showed minimal differences in health outcomes between two forms of hazelnuts (raw versus roasted/lightly salted) [35].
- Avoid, or decrease the consumption of those nuts commercialized with salt, honey, chocolate, etc.
- Eat them with the skin, pellicle, or outer soft shell, rich in vitamins and antioxidants.
- To guarantee their freshness, if we remove their nutshell, it is convenient to preserve them in a hermetic recipient in the refrigerator.
- If we buy nuts with nutshells, nutshells should be clean and without cracks. If they are packaging without skin, we should verify the absence of mold.

5.3.2 How to Introduce Nuts in Our Diets

Nuts are very versatile and they combine very well with a lot of food products. Therefore, to introduce them into our daily diet does not mean a great effort.

- We can eat them as a whole, hash, or powered (if there is any difficulty to chew or swallow adequately).

- It is possible to consume them once in a time or distributed along the day.
- For children, it is desirable to provide them as powered to avoid chokes, and in little amounts to avoid food illness.
- They are excellent as snacks during midmorning or midafternoon. Their satiety effect will help us to decrease our hungry feeling.
- We can include them in the breakfast with cereals or putting them on a toast of bread.
- If we include them into the salads, we will obtain a more savory and nutritious dish. We can use them to cook dressings.
- They provide a pleasant taste to the vegetables, rice, and soups dishes.
- The pesto sauce, elaborated with pine nuts, is delicious as a dressing for pasta; however, any kind of nuts is a good accompaniment.
- Hash or powered in yogurts or other dairy products are a good dessert.
- House made confectionary also admits nuts: cookies, biscuits, cakes, puddings, etc.
- Hash with a little olive oil or cheese is an excellent substitute to spread on bread instead of butter or margarine.
- As a sauce ingredient for meats and fish.
- We can use them in buttered or breaded food.
- As a dessert, with cheese or fruits: fruit salads, roasted apples stuffed with nuts, milkshakes, etc.

Because of the available scientific evidence, leading experts in nutrition have recommended nuts as a component of healthy dietary patterns [36].

The 2015 Dietary Guidelines for Americans recommends the consumption of 5-oz equivalent per week of nuts, seeds, and soy products. Nuts are also included in the American Heart Association recommendations and the U.S. Food and Drug Administration allows the claim "Eating a diet that includes one ounce of nuts daily can reduce your risk of heart disease."

REFERENCES

[1] Sabate J, Salas-Salvado J, Ros E. Nuts: nutrition and health outcomes. Br J Nutr 2006;96 (Suppl. 2):1S−102S.
[2] Fraser GE, Sabate J, Beeson WL, Strahan TM. A possible protective effect of nut consumption on risk of coronary heart disease: the Adventist Health Study. Arch Intern Med 1992;152:1416−24.
[3] Sabate J, Fraser GE, Burke K, Knutsen SF, Bennett H, Lindsted KD. Effects of walnuts on serum lipid levels and blood pressure in normal men. N Engl J Med 1993;328:603−7.
[4] Ros E, Tapsell LC, Sabate J. Nuts and berries for heart health. Curr Atheroscler Rep 2010;12:397−406.
[5] Bach-Faig A, Berry EM, Lairon D, Reguant J, Trichopoulou A, Dernini S, et al. Mediterranean Diet Foundation Expert Group: Mediterranean diet pyramid today. Public Health Nutr 2011;14:2274−84.

[6] Trichopoulou A, Martínez-González MA, Tong TY, Forouhi NG, Khandelwal S, Prabhakaran D, et al. Definitions and potential health benefits of the Mediterranean diet: views from experts around the world. BMC Med 2014;12:112.

[7] Josse AR, Kendall CV, Augustin LS, Ellis PR, Jenkins DJ. Almonds and postprandial glycemia—a dose—response study. Metabolism 2007;56:400—4.

[8] Ros E. Nuts and novel biomarkers of cardiovascular disease. Am J Clin Nutr 2009;89:1649S—1656SS.

[9] Ros E. Health benefits of nut consumption. Nutrients 2010;2:652—82.

[10] Luo C, Zhang Y, Ding Y, Shan Z, Chen S, Yu M, et al. Nut consumption and risk of type 2 diabetes, cardiovascular disease, and all-cause mortality: a systematic review and meta-analysis. Am J Clin Nutr 2014;100:256—69.

[11] Yu Z, Malik VS, Keum N, Hu FB, Giovannucci EL, Stampfer MJ, et al. Associations between nut consumption and inflammatory biomarkers. Am J Clin Nutr 2016;104:722—8.

[12] Afshin A, Micha R, Khatibzadeh S, Mozaffarian D. Consumption of nuts and legumes and risk of incident ischemic heart disease, stroke, and diabetes: a systematic review and meta-analysis. Am J Clin Nutr 2014;100:277—88.

[13] Aune D, Keum N, Giovannucci E, Fadnes LT, Boffetta P, Greenwood DC, et al. Nut consumption and risk of cardiovascular disease, total cancer, all-cause and cause-specific mortality: a systmatic review and dose-response meta-analysis of prospective studies. BMC Medicine 2016;14:207.

[14] Eslamparast T, Sharafkhah M, Poustchi H, Hashemian M, Dawsey SM, Freedman ND, et al. Nut consumption and total and cause-specific mortality: results from the Golestan Cohort Study. Int J Epidemiol 2016; pii: dyv365. [Epub ahead of print].

[15] O'Neil CE, Keast DR, Nicklas TA, Fulgoni VL. Nut consumption is associated with decreased health risk factors for cardiovascular disease and metabolic syndrome in U.S. adults: NHANES 1999-2004. J Am Coll Nutr 2011;30:502—10.

[16] Ibarrola-Jurado N, Bullo M, Guasch-Ferre M, Ros E, Martinez-Gonzalez MA, Corella D, et al. PREDIMED Study Investigators. Cross-sectional assessment of nut consumption and obesity, metabolic syndrome and other cardiometabolic risk factors: the PREDIMED study. PLoS One 2013;8:e57367.

[17] Fernandez-Montero A, Bes-Rastrollo M, Beunza JJ, Barrio-Lopez MT, de la Fuente-Arrillaga C, Moreno-Galarraga L, et al. Nut consumption and incidence of metabolic syndrome after 6-year follow-up: the SUN (Seguimiento Universidad de Navarra, University of Navarra Follow-up) cohort. Public Health Nutr 2013;16:2064—72.

[18] Salas-Salvado J, Guasch-Ferre M, Bullo M, Sabate J. Nuts in the prevention and treatment of metabolic syndrome. Am J Clin Nutr 2014;100(Suppl):399S—407S.

[19] Sabate J, Oda K, Ros E. Nut consumption and blood lipid levels: a pooled analysis of 25 intervention trials. Arch Intern Med 2010;170:821—7.

[20] Banel DK, Hu FB. Effects of walnut consumption on blood lipids and other cardiovascular risk factors: a meta-analysis and systematic review. Am J Clin Nutr 2009;90:56—63.

[21] Estruch R, Ros E, Salas-Salvado J, Covas MI, Corella D, Aros F, et al. PREDIMED Study Investigators. Primary prevention of cardiovascular disease with a Mediterranean diet. N Engl J Med 2013;368:1279—90.

[22] Estruch R, Martinez-Gonzalez MA, Corella D, Salas-Salvado J, Ruiz-Gutierrez V, Covas MI, et al. PREDIMED Study Investigators. Effects of a Mediterranean-style diet on cardiovascular risk factors: a randomized trial. Ann Intern Med 2006;145:1—11.

[23] Salas-Salvado J, Fernandez-Ballart J, Ros E, Martinez-Gonzalez MA, Fito M, Estruch R, et al. PREDIMED Study Investigators. Effect of a Mediterranean diet supplemented with nuts on metabolic syndrome status: one-year results of the PREDIMED randomized trial. Arch Intern Med 2008;168:2449–58.

[24] Salas-Salvado J, Bullo M, Estruch R, Ros E, Covas MI, Ibarrola-Jurado N, et al. Prevention of diabetes with Mediterranean diets. A subgroup analysis of a randomized trial. Ann Intern Med 2014;160:1–10.

[25] Ruiz-Canela M, Estruch R, Corella D, Salas-Salvadó J, Martínez-González MA. Association of a Mediterranean diet with peripheral artery disease: the PREDIMED randomized trial. JAMA 2014;311:415–17.

[26] Valls-Pedret C, Sala-Vila A, Serra-Mir M, Corella D, de la Torre R, Martinez-Gonzalez MA, et al. Mediterranean Diet and age-related cognitive decline: a randomized clinical trial. JAMA Intern Med 2015;175:1094–103.

[27] Martinez-Gonzalez MA, Bes-Rastrollo M. Nut consumption, weight gain and obesity: epidemiological evidence. Nutr Metab Cardiovasc Dis 2011;21:S40–5.

[28] Jackson CL, Hu FB. Long-term associations of nut consumption with body weight and obesity. Am J Clin Nutr 2014;100(Suppl):408S–11SS.

[29] Mattes RD, Kris-Etherton PM, Foster GD. Impact of peanuts and tree nuts on body weight and healthy weight loss in adults. J Nutr 2008;138:1741S–5SS.

[30] Ellis PR, Kendall CW, Ren Y, Parker C, Pacy JF, Waldron KW, et al. Role of cell walls in the bioaccessibility of lipids in almond seeds. Am J Clin Nutr 2004;80:604–13.

[31] Bes-Rastrollo M, Sabate J, Gomez-Gracia E, Alonso A, Martinez JA, Martinez-Gonzalez MA. Nut consumption and weight gain in a Mediterranean cohort: the SUN study. Obesity, 15. Silver Spring; 2007. p. 107–16.

[32] Bes-Rastrollo M, Wedick NM, Martinez-Gonzalez MA, Li TY, Sampson L, Hu FB. Prospective study of nut consumption, long-term weight change, and obesity risk in women. Am J Clin Nutr 2009;89:1913–19.

[33] Flores-Mateo G, Rojas-Rueda D, Basora J, Ros E, Salas-Salvado J. Nut intake and adiposity: meta-analysis of clinical trials. Am J Clin Nutr 2013;97:1346–55.

[34] Estruch R, Martinez-Gonzalez MA, Corella D, Salas-Salvado J, Fito M, Chiva-Blanch G, et al. Effect of a high-fat Mediterranean diet on body weight and waist circumference: a prespecified secondary outcomes analysis of the PREDIMED randomised controlled trial. Lancet Diabetes Endocrinol 2016;4:666–76.

[35] Tey SL, Robinson T, Gray AR, Chisholm AW, Brown RC. Do dry roasting, lightly salting nuts affect their cardioprotective properties and acceptability? Eur J Nutr 2016;56 (3):1025–36. [Epub ahead of print].

[36] Hu FB, Willet WC. Optimal diets for prevention of coronary heart disease. JAMA 2002;288:2569–78.

Chapter 6

Fruits and Vegetables

Angeliki Papadaki and Ana Sánchez-Tainta

6.1 INTRODUCTION

Fruits and vegetables have always formed an important part of the traditional Mediterranean Diet. In the early 1960s, the Seven Countries Study showed that people in the Greek island of Crete consumed approximately 654 g of fruits and vegetables every day (approximately eight servings per day) [1,2]. This was because every meal was accompanied by a salad of either raw or cooked vegetables. Vegetables were regularly consumed as a main dish, like stews and casseroles, and they were also regularly added to meat dishes, like stews (e.g., chicken with okra, peas or green beans, cooked in a rich tomato sauce sautéed in olive oil and onions). Fruit was the main dessert in the traditional Mediterranean Diet, eaten not only after every main meal but also consumed between meals, as a snack.

It is now recognized that the low chronic disease, including cardiovascular disease, rates and high life expectancy of Cretan people in the 1960s were attributed to the variety of the Cretan diet instead of a specific food or nutrient as well as to lifestyle factors such as physical activity and the fact that meals were an opportunity of socializing with family and friends and thus, provided relief and relaxation from daily stress [3]. Being a frequent food of this dietary pattern, however, fruits and vegetables provided many of the known essential nutrients and plant substances believed to promote health and prevent cardiovascular disease [4,5]. In particular, fruits and vegetables are rich in dietary fiber, vitamins (e.g., vitamin C, carotenes, and folate), minerals (e.g., potassium and calcium), and phytochemicals (e.g., flavonoids and anthocyanins), which are bioactive compounds with antioxidant abilities and other beneficial properties that provide cardiovascular protection. The beneficial role of fruits and vegetables in cardiovascular health is attributed to this combination of nutrients and nonnutrient compounds that are abundant in these foods [6].

The Prevention of Cardiovascular Disease through The Mediterranean Diet.
DOI: http://dx.doi.org/10.1016/B978-0-12-811259-5.00006-8

6.2   CONSUMPTION OF FRUITS AND VEGETABLES AND RISK OF CARDIOVASCULAR DISEASE

Fruits and vegetables have been suggested to protect against cardiovascular disease via several biological mechanisms that involve the antioxidant, antiinflammatory and antiplatelet properties of their constituents as well as their ability to inhibit cholesterol synthesis and control blood pressure levels. For example, carotenes such as lycopene and anthocyanins may protect against cardiovascular disease by preventing the formation of clots [7] and inhibiting the synthesis of cholesterol in the body [8], whereas carotenes such as beta-carotene and beta-cryptoxanthin have antiinflammatory properties [9]. Vitamin C and flavonoids, such as quercetin, act as antioxidants by helping prevent free radical formation and reducing oxidative stress [6,10] and sulfides, such as allicin, may prevent heart disease by lowering cholesterol and blood pressure levels [11]. In addition, indoles and isothiocyanates, compounds often found in green fruits and vegetables, are known for their antioxidant, antiplatelet, antithrombotic, and antiinflammatory properties [6,12,13]. Fruits and vegetables are also naturally (i.e., in an unprocessed form) low in sodium and a good source of potassium as well as low in energy density, thereby potentially playing a role in blood pressure control by reducing platelet aggregation and thrombosis, thereby diminishing the rate of atherosclerotic lesion formation as well as contributing to body weight control [14]. Dietary fiber found in fruits and vegetables has been associated with lower risk of cardiovascular disease via mechanisms involving the reduction in blood cholesterol and glucose levels and induced feelings of satiety, thus contributing to reduced body weight gain [15].

In a recent systematic review and meta-analysis, Wang et al. examined the dose-response association between consumption of fruits and vegetables and mortality from cardiovascular disease [16]. The findings suggested moderate evidence of a reduction in the risk of cardiovascular mortality of 4% for every additional serving of fruits and vegetables consumed per day. When fruits and vegetables were examined separately, there was moderate evidence of a 5% cardiovascular risk reduction for every additional serving of fruit, and 4% reduction in cardiovascular mortality risk for every additional serving of vegetable consumed per day [16].

When considering individual types of cardiovascular disease, two types that have been extensively examined are coronary heart disease and stroke. In their systematic review and meta-analysis, Dauchet et al. examined the association between fruit and vegetable consumption and coronary heart disease [17]. The findings showed strong evidence that coronary heart disease risk decreased by 4% for each additional serving of fruits and vegetables consumed per day and by 7% for each additional serving of fruit consumed on a daily basis. In addition, there was strong evidence that increases in daily vegetable intake by one serving were associated with a

26% reduction in cardiovascular mortality [17]. However, the authors suggested that publication bias might have affected the results of their review, thus potentially overestimating the estimated risk reduction, and concluded that experimental studies would be needed to confirm these findings, in addition to explaining these associations.

A systematic review and meta-analysis of nine prospective studies examined the association between fruit and vegetable consumption and the risk of stroke [18]. Strong evidence was found that, compared to a daily consumption of less than three servings of fruits and vegetables, consuming 3−5 servings on a daily basis is associated with a reduction in the risk of stroke by 11% and consuming more than five servings per day is associated with a reduction of 26%. These associations were observed for both ischemic and hemorrhagic stroke [18]. A more recent meta-analysis of 20 prospective cohort studies showed that higher, versus lower, fruit and vegetable consumption was associated with a 21% reduction in the risk of stroke, whereas higher fruit intake and vegetable intakes were associated with a 23% and 14% risk reduction, respectively [19]. In subgroup analyses, citrus fruit, leafy vegetables, and apples/pears seemed to contribute more to this protective association, compared to other fruit and vegetable varieties.

6.3 CONSUMPTION OF FRUITS AND VEGETABLES AND CARDIOVASCULAR RISK FACTORS

6.3.1 Obesity

Fruits and vegetables have been suggested to constitute a potential tool in the prevention and/or management of obesity, due to their high water and dietary fiber, and low energy content [20]. Dietary fiber in fruits and vegetables have a favorable impact on satiety and slow gastric emptying, thereby creating a feeling of fullness that prevents overeating and results in lower energy intake [21]. In addition, many fruits and vegetables are low in glycemic index and glycemic load, thereby preventing postprandial hormonal changes that might increase appetite and energy intake [22]. Also, some research has suggested that fruits and vegetables can protect against obesity because they replace more energy-dense foods or snacks in the diet [23].

The role of fruits and vegetables in the prevention and/or management of obesity remains unclear, however. In their systematic review, Ledoux et al. investigated the relationship between consumption of fruits and vegetables with adiposity [24]. They found that higher fruit and vegetable consumption is associated with slower weight gain in overweight adults, but this association appears to be weak. Intervention studies included in this review indicated that increased fruit and vegetable consumption contributes to reduced adiposity in overweight and/or obese adults, but not children. The authors concluded that more research is needed to elucidate the role of fruits and vegetables, in isolation from other dietary factors, as well as their potential mechanisms of action in the management of obesity [24].

In a more recent systematic review, Kaiser et al. examined the effect of increased consumption of fruits and vegetables on weight loss or the prevention of weight gain. Their findings showed that recommendations to increase fruit and vegetable consumption do not lead to meaningful weight loss (mean change = −0.16 kg) [25]. The authors concluded that recommending an increased consumption of fruits and vegetables alone, particularly without promoting reduction of total energy intake, does not support the notion that these foods can protect against obesity [25]. In another systematic review examining the effect of increased fruit and vegetable consumption on body weight and energy intake, Mytton et al. found moderate evidence for a difference in body weight change of 0.68 kg between higher, compared to lower, fruit and vegetable intake [26]. There was weak evidence for a difference in total energy intake changes between high and low intake. The authors concluded that promoting an increased consumption of fruits and vegetables, without recommending for them to replace other foods, will unlikely lead to weight gain in the short term (up to a year) [26].

Since obesity contributes to cardiovascular disease and other cardiovascular risk factors, such as increased blood pressure and type 2 diabetes, and because fruits and vegetables do not seem to cause weight gain, their consumption, as part of a balanced Mediterranean diet and according to dietary recommendations, can aid the prevention of cardiovascular disease.

6.3.2   Hypertension

Fruits and vegetables have been suggested to play a role in hypertension prevention and management, due to their low sodium content and their high content of potassium, several vitamins, flavonoids and carotenoids, which have been implicated in blood pressure reductions via improving endothelial function and vasodilation and increasing antioxidant activity [27,28]. In a recent systematic review, Li et al. examined the association between fruit and vegetable consumption and risk of hypertension [29]. The authors found strong evidence that hypertension risk was reduced by 19% when comparing the highest with the lowest total fruit and vegetable consumption and 27% when comparing the highest with the lowest fruit consumption. There was weak evidence for a hypertension risk reduction, of 3%, between the highest and lowest vegetable intake. The authors concluded that fruit and vegetable consumption might be inversely associated with risk of hypertension, but more research is needed in this area due to the different methods employed in the studies reviewed (e.g., the observed weak evidence of an association between vegetable intake and hypertension risk might be due to differences in vegetable preparation, preservation and/or cooking methods) [29].

6.3.3 Type 2 Diabetes

Fruits and vegetables have been suggested to protect against the risk of developing type 2 diabetes due to several potential mechanisms. Most fruits and vegetables have a low glycemic index and load, which can help control blood glucose and insulin levels and improve glucose intolerance [30]. The high phytochemical content of fruits and vegetables has also been implicated in the protection against type 2 diabetes, due to the antioxidant properties they exert [31]. Indirectly, fruit and vegetable consumption might protect against type 2 diabetes through mechanisms associated with energy intake and body weight [32].

A recent systematic review aimed to investigate the association between fruit and vegetable consumption and incidence of type 2 diabetes [33]. There was weak evidence of an association between increased vegetables, fruit, or fruits and vegetables combined, but there was moderate evidence of a reduction (14%) in type 2 diabetes incidence with greater consumption of green leafy vegetables. The authors concluded that increasing the consumption of green leafy vegetables is inversely associated with type 2 diabetes incidence and that further research to elucidate this association is needed [33]. This finding is in agreement with another systematic review and meta-analysis that examined the dose-response association between fruits and vegetables and type 2 diabetes risk [34]. Similar to the review by Carter et al. [33], Li et al. found that there was moderate evidence of a reduction in type 2 diabetes risk of 13% for an increase of 0.2 servings per day of green leafy vegetables, and of a 7% reduction for an increase of one serving of fruit per day, whereas there was weak evidence of risk reduction (10%) for an increase of one serving of vegetables per day [34]. The evidence of an association between the consumption of fruits and vegetables (combined) and type 2 diabetes risk was weak.

Type 2 diabetes is an independent risk factor for cardiovascular disease, and the consumption of fruits and vegetables, containing a variety of phytochemicals, might play a role in diabetes prevention due to their antioxidant properties. Although results from systematic reviews seem inconclusive, it appears that green leafy vegetables, in particular, have a protective role in type 2 diabetes prevention and should be consumed as part of a balanced Mediterranean Diet.

6.4 HOW MUCH SHOULD I EAT?

Eating fruit and vegetables every day, as part of a Mediterranean-style diet, is one of the most important things you can do to prevent cardiovascular disease. The Dietary Guidelines for Americans (2010) recommended that we should consume at least 9 servings of fruits and vegetables (five servings of vegetables and four servings of fruit) every day [35]. This is very close to the amount of fruits and vegetables consumed in the traditional diet of Crete [1,2]. The most recent guidelines for the American population from 2015 also focus on the importance of consuming a variety of vegetables with different colors and eating whole fruits [36].

6.5   WHAT COUNTS AS A SERVING?

An average serving of fruit is:

- 1 medium-size piece of fruit (apple, banana, pear, orange, peach)
- 2 small fruits (plums, kiwis)
- ½ grapefruit
- 1 slice of large fruit (melon, watermelon, pineapple)
- ½ −1 tablespoon of dried fruit (raisins, apricots)
- 1 handful of grapes, cherries, berries, strawberries
- 2−3 tablespoons of fruit salad or canned fruit (choose the varieties canned in natural fruit juice, instead of syrup)

An average serving of vegetables is:

- ½ cup cooked vegetables
- 1 cup raw vegetables

Although the traditional Mediterranean Diet was characterized by minimal processing and a high proportion of fresh, seasonal fruits and vegetables, all varieties count toward your day-to-day intake (fresh, dried, frozen, and canned) [37]. There are also so many different types, that you can increase your consumption without getting bored. Eating different types of fruits and vegetables every day is also important because every kind has different types of phytochemicals and each phytochemical protects you in a different way. One easy way to do this is to think that different phytochemicals are found in different colors of fruits and vegetables. Think of:

- Red fruits and vegetables (e.g., tomatoes, beetroot, onions, peppers, strawberries, watermelon, grapes, apples) rich in carotenes, such as lycopene, and anthocyanins.
- White fruits and vegetables (e.g., onions, leeks, mushrooms, cauliflower, bananas, pears) rich in flavonoids, such as quercetin, and sulfides, such as allicin.
- Green fruits and vegetables (e.g., spinach, broccoli, kale, collard greens, kiwi fruit, apples, grapes, pears) rich in indoles and isothiocyanates.
- Blue/purple fruits and vegetables (e.g., blueberries, plums, black currants, blackberries, cabbage, eggplants) rich in anthocyanins.
- Orange/yellow fruits and vegetables (e.g., carrots, peppers, pumpkins, butternut squash, oranges, apricots, nectarines, mangos) rich in carotenes, such as beta-carotene and beta-cryptoxanthin.

6.6 HOW CAN I INCREASE MY FRUIT AND VEGETABLE CONSUMPTION?

Eating **at least** nine servings of fruits and vegetables every day can be easily achieved. The following tips will help you:

- Keep a bowl of fresh fruit on your kitchen counter. This will make it more likely for you to grab and eat when you want a quick snack.
- Add fresh, canned or dried fruit to your breakfast cereal.
- When eating out, choose fruit-based desserts or fruit salads topped with yoghurt.
- Add pieces of fruit to your salad, for an even crunchier texture, variety, and color. Apples, oranges, pineapple, and raisins are ideal add-ins to salads.
- Pick up a variety of fresh fruit when it's your turn to buy your colleagues the coffee-break snacks.
- Add vegetables to your sandwiches, pizzas, hamburgers, fajitas, and casseroles. Try lettuce, onions, mushrooms, cucumber, tomato, peppers, or any other vegetable you like.
- Have a green salad with every main meal. Give it more color by adding peppers, cucumber, and tomatoes. Fill half of your plate with your salad first, and then add other foods. Ask for vegetables on the side, or a side salad, when you eat out.
- Prepare chopped vegetables, like carrots, broccoli, cauliflower, celery, peppers, and zucchini. Keep them in a container sprinkled with lemon juice. Munch with no guilt or enjoy them with low-fat dips when feeling hungry between meals.
- Take advantage of the ready-cut vegetables and salads, if you don't have the time to prepare your own salads.
- Make vegetables visible in your refrigerator or kitchen counter. The more often you see them, the more likely you'll eat them.

Set out to meet most of these goals. Start slowly and try to increase the amount of fruits and vegetables you eat from there. Sooner than you think, these tips will become part of your everyday life.

6.7 SUMMARY AND RECOMMENDATIONS

The frequent consumption of fruits and vegetables is one of the main characteristics of the traditional Mediterranean Diet [4]. The beneficial role of fruits and vegetables on cardiovascular disease has been demonstrated in many prospective studies. Moreover, fruits and vegetables are rich in dietary fiber, vitamins, and phytochemicals, and low in energy density and glycemic index (most fruits and vegetables), which might explain these beneficial associations. Because these compounds are numerous, complex and their interactions are not completely understood, recommendations emphasize the importance of consuming a variety of plant foods, including fruits and vegetables, instead of single nutrients. Fruits and vegetables, however, although abundant in the traditional Mediterranean Diet, are not consumed in optimal amounts in current Western diets. A gradual incorporation of fruits and vegetables, as part of a health-promoting diet, however, could be achieved without difficulty. The benefits of even a moderate consumption in reducing cardiovascular disease could far outweigh any restraints due to potential lack of familiarity of tastes, cost, or lack of cooking skills.

REFERENCES

[1] Ferro-Luzzi A, Branca F. Mediterranean diet, Italian-style: prototype of a healthy diet. Am J Clin Nutr 1995;61(6):1338S−45S.

[2] Kushi LH, Lenart EB, Willett WC. Health implications of Mediterranean diets in light of contemporary knowledge. 2. Meat, wine, fats and oils. Am J Clin Nutr 1995;61 (6):1416S−27S.

[3] Willett WC, Sacks F, Trichopoulou A, Drescher G, Ferro-Luzzi A, Helsing E, et al. Mediterranean diet pyramid: a cultural model for healthy eating. Am J Clin Nutr 1995;61 (6):1402S−6S.

[4] Kafatos A, Verhagen H, Moschandreas J, Apostolaki I, Van Westerop JJM. Mediterranean diet of Crete: foods and nutrient content. J Am Diet Assoc 2000;100(12):1487−93.

[5] Willett WC. Diet and health: What should we eat? Science 1994;264(5158):532−7.

[6] Liu RH. Health promoting components of fruits and vegetables in the diet. Adv Nutr 2013;4:384S−92S.

[7] Krugera MJ, Daviesa N, Myburghb KH, Lecourc S. Proanthocyanidins, anthocyanins and cardiovascular diseases. Food Res Int 2014;59:41−52.

[8] Arab L, Steck S. Lycopene and cardiovascular disease. Am J Clin Nutr 2000;71 (6):1691S−5S.

[9] Kritchevsky SB. β-Carotene, carotenoids and the prevention of coronary heart disease. J Nutr 1999;129(1):5−8.

[10] Perez-Vizcaino F, Duarte J, Andriantsitohaina R. Endothelial function and cardiovascular disease: effects of quercetin and wine polyphenols. Free Radic Res 2006;40(10):1054−65.

[11] Bradley JM, Organ CL, Lefer DJ. Garlic-derived organic polysulfides and myocardial protection. J Nutr 2016;146(2):403S−9S.

[12] Dinkova-Kostova AT, Kostov RV. Glucosinolates and isothiocyanates in health and disease. Trends Mol Med 2012;18(6):337−47.

[13] Park MK, Rhee YH, Lee HJ, Lee EO, Kim KH, Park MJ, et al. Antiplatelet and antithrombotic activity of indole-3-carbinol in vitro and in vivo. Phytother Res 2008;22(1):58−64.

[14] Slavin JL, Lloyd B. Health benefits of fruits and vegetables. Adv Nutr 2012;3:506−16.

[15] Threapleton DE, Greenwood DC, Evans CEL, Cleghorn CL, Nykjaer C, Woodhead C, et al. Dietary fibre intake and risk of cardiovascular disease: systematic review and meta-analysis. BMJ 2013;347:f6879.

[16] Wang X, Ouyang Y, Liu J, Zhu M, Zhao G, Bao W, et al. Fruit and vegetable consumption and mortality from all causes, cardiovascular disease, and cancer: systematic review and dose-response meta-analysis of prospective cohort studies. BMJ 2014;349:g4490.

[17] Dauchet L, Amouyel P, Hercberg S, Dallongeville J. Fruit and vegetable consumption and risk of coronary heart disease: a meta-analysis of cohort studies. J Nutr 2006;136:2588−93.

[18] He FJ, Nowson CA, MacGregor GA. Fruit and vegetable consumption and stroke: meta-analysis of cohort studies. Lancet 2006;367:329-6.

[19] Hu D, Huang J, Wang Y, Zhang D, Qu Y. Fruits and vegetables consumption and risk of stroke: a meta-analysis of prospective cohort studies. Stroke 2014;45:1613−19.

[20] World Health Organization. Diet, nutrition and the prevention of chronic diseases: report of the joint WHO/FAO expert consultation: WHO technical report series No. 916 (TRS 916), Geneva; 2002.

[21] Howarth NC, Saltzman E, Roberts SB. Dietary fiber and weight regulation. Nutr Rev 2001;59:129−39.

[22] Livesey G, Taylor R, Hulshof T, Howlett J. Glycemic response and health—a systematic review and meta-analysis: relations between dietary glycemic properties and health outcomes. Am J Clin Nutr 2008;87:258S−68S.

[23] Rolls BJ, Ello-Martin JA, Tohill BC. What can intervention studies tell us about the relationship between fruit and vegetable consumption and weight management? Nutr Rev 2004;62:1−17.

[24] Ledoux TA, Hingle MD, Baranowski T. Relationship of fruit and vegetable intake with adiposity: a systematic review. Obes Rev 2011;12:e143−50.

[25] Kaiser KA, Brown AW, Bohan Brown MM, Shikany JM, Mattes RD, Allison DB. Increased fruit and vegetable intake has no discernible effect on weight loss: a systematic review and meta-analysis. Am J Clin Nutr 2014;100:567−76.

[26] Mytton OT, Nnoaham K, Eyles H, Scarborough P, Ni Mhurchu C. Systematic review and meta-analysis of the effect of increased vegetable and fruit consumption on body weight and energy intake. BMC Public Health 2014;14:886.

[27] Toh JY, Tan VM, Lim PC, Lim ST, Chong MF. Flavonoids from fruit and vegetables: a focus on cardiovascular risk factors. Curr Atheroscler Rep 2013;15:368.

[28] Aburto NJ, Hanson S, Gutierrez H, Hooper L, Elliott P, Cappuccio FP. Effect of increased potassium intake on cardiovascular risk factors and disease: systematic review and meta-analyses. BMJ 2013;346:f1378.

[29] Li B, Li F, Wang L, Zhang D. Fruit and vegetables consumption and risk of hypertension: a meta-analysis. J Clin Hypertens (Greenwich) 2016;18:468−76.

[30] Greenwood DC, Threapleton DE, Evans CE, Cleghorn CL, Nykjaer C, Woodhead C, et al. Glycemic index, glycemic load, carbohydrates, and type 2 diabetes: systematic review and dose-response meta-analysis of prospective studies. Diabetes Care 2013;36(12):4166−71.

[31] Liu RH. Health benefits of fruit and vegetables are from additive and synergistic combinations of phytochemicals. Am J Clin Nutr 2003;78:517S−20S.

[32] Mozaffarian D, Hao T, Rimm EB, Willett WC, Hu FB. Changes in diet and lifestyle and long-term weight gain in women and men. N Engl J Med 2011;364:2392−404.

[33] Carter P, Gray LJ, Troughton J, Khunti K, Davies MJ. Fruit and vegetable intake and incidence of type 2 diabetes mellitus: systematic review and meta-analysis. BMJ 2010;341: c4229.

[34] Li M, Fan Y, Zhang X, Hou W, Tang Z. Fruit and vegetable intake and risk of type 2 diabetes mellitus: meta-analysis of prospective cohort studies. BMJ Open 2014;4:e005497.

[35] USDA. Dietary Guidelines for Americans 2010. Hyattsville, MD: USDA Human Nutrition Information Service; 2010.

[36] U.S. Department of Health and Human Services, U.S. Department of Agriculture. 2015−2020 Dietary Guidelines for Americans. <http://health.gov/dietaryguidelines/2015/guidelines/>; 2015.

[37] Barrett DM. Maximizing the nutritional value of fruits and vegetables. Food Technol 2007;61(4):40−4.

Chapter 7

Cereals and Legumes

Karen J. Murphy, Iva Marques-Lopes and Ana Sánchez-Tainta

7.1 CEREALS AND CARDIOVASCULAR DISEASE

7.1.1 Overview and Introduction of Cereals in the Mediterranean Diet

Cereals include foods and food products including rice, oats, quinoa, pasta, bread, cereals, crackers, etc. Global dietary recommendations generally suggest the inclusion of cereals as mainly whole-grain cereals as part of a healthy eating pattern. Cereals in general provide valuable nutrients such as B vitamins; vitamin A; minerals including iron, zinc, and selenium; and dietary fiber; however processing cereals, which removes the bran and germ layers, often results in depletion of fiber and nutrients. Many refined cereal products result in high levels of added sugars, sodium, and saturated fat and require the addition of vitamins and minerals back into the product to replace what was lost during processing.

Cereals in the Mediterranean Diet include bread, pasta, rice, couscous, and other mainly whole-grain cereals [1,2]. Revisions of the Mediterranean diet pyramids have seen changes to recommendations for cereals, from eight serves of breads and cereals daily (The 1999 Greek Dietary Guidelines), to breads and cereals at every meal (Oldway's preservation and Trust 2009 pyramid) to the current recommendations of one to two serves of breads, rice, pasta, couscous, and other whole-grain cereals at every meal [1]. Consumption of whole-grain cereals over refined cereals are generally recommended for the prevention of CVD as they contain minerals, antioxidants, fiber, starch, and phytonutrients, all reported to have some cardioprotection [3]. The Mediterranean dietary pattern represents a healthy profile of dietary fat, a low proportion of refined cereals but high in whole-grain cereals, a low glycemic index, and is low in discretionary foods. This pattern delivers dietary fiber, vitamins, minerals, and antioxidant and antiinflammatory compounds, and has been shown to reduce the risk of mortality, CVD-related mortality as well as type 2 diabetes mellitus, CVD risk factors, metabolic syndrome, and several other chronic diseases [4–8].

The Prevention of Cardiovascular Disease through The Mediterranean Diet.
DOI: http://dx.doi.org/10.1016/B978-0-12-811259-5.00007-X

When studying dietary patterns it is difficult to associate an effect with a specific dietary component. It may be the complete dietary pattern and synergy between foods and nutrients responsible for positive health outcomes. Nevertheless, it is important to understand the health benefits that single foods offer and when consumed together can provide an indication of overall dietary quality. We know from research that high adherence to the Mediterranean Diet is associated with reduced all-cause mortality, CVD-related mortality, incident type 2 diabetes mellitus as well as reduced inflammatory markers, homocysteine levels, dyslipidemia and hypertension, and other conditions such as Alzheimer's disease [4−9]. In the PREDIMED trial (Prevención con Dieta Mediterránea in Spanish), a Mediterranean Diet supplemented with nuts or extra-virgin olive oil were compared with a low-fat diet in 7447 high CVD risk individuals with a follow-up period of ∼5 years. The authors showed a 30% risk reduction in cardiovascular events in the Mediterranean Diet groups compared with the low-fat diet group [4]. When diabetes-free participants were selected and the two Mediterranean Diet groups pooled, compared with the low-fat group type 2 diabetes incidence was reduced by 52%, in the absence of significant changes in body weight or physical activity [7]. In the PREDIMED study, the authors followed up 7216 men and women at high CVD-risk for a mean of 5.9 years. Baseline fiber intake was significantly associated with lower risk of mortality. The group with the highest dietary fiber intake consumed more fruit, vegetables, whole-grain foods, and less refined grains. The major whole-grain foods were whole-grain bread, white bread and pasta [10]. Further, in a subsample of the PREDIMED trial ($n = 772$, 69 ± 5 years), the authors assessed the impact of dietary fiber intakes on cardiovascular risk factors over a 3-month period. The high-risk men and women were allocated either a Mediterranean Diet with either nuts or extra-virgin olive oil or a low-fat diet group and received behavior and dietary education around increasing intakes of vegetables, fruits, and legumes. Blood pressure, body weight, waist circumference all decreased across quintiles of fiber intake ($P < 0.005$), while reductions in fasting glucose, total cholesterol ($P = 0.04$) and increases in HDL cholesterol ($P = 0.02$) were greatest among participants in the top 20% of fiber intake. Similarly, in participants with the largest increases in consumption of soluble fiber intake, they showed significant reductions in LDL cholesterol ($P = 0.04$).

7.1.2 Whole-Grain Cereals and Cardiovascular Disease: Epidemiological Evidence

Some systematic review and meta-analyses of the literature provide mixed or inconclusive evidence around the cardioprotective properties of whole-grain foods, however can often be explained by poorly designed, insufficiently powered studies and the use of limited whole-grain foods. Further, many studies have been funded by commercial industry with a stake in whole grains which also needs to be considered when evaluating results. However, more recently, evaluation of prospective data in over 2.7 million individuals has shown a clear association with high whole-grain consumption and lower total and CVD-related mortality [11]. Furthermore, dietary patterns rich in whole grains have been associated with reduced risk of coronary heart disease and type 2 diabetes mellitus, independent of individual nutrients found in whole grains [12].

Whole-grain consumption is associated with improvements in individual risk factors for CVD including body mass and abdominal obesity [13], insulin resistance and insulin sensitivity [14,15], C-reactive protein [3], blood lipids [16,17], and blood pressure [18,19]. Whole-grain cereals have greater energy density and purported greater satiation than that of refined cereals, which may help explain their positive effect on weight control [13]. In the Framingham Offspring Study [12], a longitudinal, community-based study of CVD among the offspring of the original participants of the Framingham Heart Study, showed an inverse association with whole grains and body mass index ($P = 0.06$) and waist to hip ratio ($P = 0.005$). A recent study by Serra-Majem and Bautista-Castano [20] showed that there is a difference between whole-grain bread and white bread and influence on body weight and abdominal fat. The author's systematic review showed that whole-grain bread does not influence weight gain however; white bread may have a possible relationship with excess abdominal fat [21]. In the PREDIMED study, participants who had the highest quartile of the change in white bread intake gained ~ 0.8 kg more and increased waist circumference ~ 1.3 cm more than those in the lowest quartile; whereas there was no significant relationship between whole-grain bread consumption and anthropometry [20]. In a weight-loss study by Katcher and colleagues [3], 50 obese men and women were prescribed a hypocaloric diet with advice on avoiding either whole-grains or consuming all grain from whole-grain sources. After 12 weeks those consuming whole grains had significantly higher intakes of dietary fiber ($P = 0.007$) and magnesium ($P < 0.001$) and lower levels of C-reactive protein, independent of weight loss, than the group avoiding whole grains ($P = 0.01$).

A systematic review from the Cochrane library [22] reviewed the literature on whole grains and total cholesterol, LDL and HDL cholesterol and triglycerides. Eight of nine studies reported on oatmeal. The weighted mean difference was -0.19 mmol/L (95% confidence interval (CI): -0.30, -0.08; $P = 0.0005$) for the oatmeal vs refined grain diets. However, because of the lack of diversity on results for other grains it is difficult to draw conclusions on the effect of whole grains, other than oatmeal on total cholesterol. This was a similar case for LDL cholesterol where the majority of studies analyzed were based on oatmeal, but still resulted in a weighted mean difference of -0.18 mmol/L (95% CI: -0.28, -0.09; $P < 0.0001$).

There was no significant effect of whole grains versus refined grain on HDL cholesterol, nor on triglycerides. Despite these inconclusive results, research from prospective studies show inverse associations with diets rich in whole grains and total and LDL cholesterol [12], however further research is needed.

While the ability of soluble fibers to reduce blood cholesterol levels is questionable, a meta-analysis of 67 controlled studies with 2990 participants [23] was undertaken to determine the relationship between soluble fibers including pectin, oat bran, guar gum, and psyllium on lowering blood cholesterol. Soluble fibre in the range of 2−10 g/d, was associated with small but significant reductions in total cholesterol (−0.045 mmol/L/g soluble fibre, 95%CI: −0.054, −0.035) and LDL cholesterol (−0.057 mmol/L/g, 95%CI: −0.070, −0.044). A high fibre diet was reported to marginally but significantly reduce HDL cholesterol (−0.002 mmol/L/g soluble fibre) but did not influence blood triglyceride levels. The mechanism by which dietary fiber may reduce cholesterol levels is not well understood. The authors suggest the mechanism may relate to binding of bile acids or cholesterol during micelle formation, leading to increased clearance of LDL cholesterol; inhibition of hepatic fatty acid synthesis; changes in intestinal motility and slower absorption of macronutrients with subsequent effects on satiety.

Evidence from dietary trials, metabolic studies and epidemiological evidence support the notion that whole-grain foods reduce diabetes risk by improving plasma glucose and insulin and subsequently glycemic control and insulin sensitivity [24]. Moreover, several dietary patterns that include whole grains, including a Mediterranean style dietary pattern are associated with reduced risk of type 2 diabetes mellitus [5]. In fact individuals who consume around three serves of whole-grain foods daily are 20%−30% less likely to develop type 2 diabetes mellitus than those who are consuming less than three serves of whole-grain foods per week [25−27]. The mechanisms thought to mitigate the risk of type 2 diabetes mellitus and help with glycemic control are not yet well understood however, several hypotheses have been raised. The physical structure of whole grains is strongly related to the reduction in risk of type 2 diabetes mellitus due to its high dietary fiber content, which may dampen the postprandial glucose and insulin responses. Moreover, the phytonutrients contained in whole-grain foods is thought to reduce oxidative stress and also offer antiinflammatory activity, both of which are implicated in the pathogenesis of insulin resistance and type 2 diabetes mellitus [24].

Epidemiological studies including the Nurses' Health Study and the Health Professionals Follow-Up Study [11], the Iowa Women's Health Study [28,29], the Atherosclerosis Risk in Communities (ARIC) Study [30], and the Framingham Offspring Study [12] have found inverse relationships with either whole-grain consumption and cereal fiber intake and risk of heart disease or total and CVD-related mortality. The Nurses' Health Study, a prospective study with 72,521 women aged 38−63 years, showed a reduced risk of CHD with high whole-grain intake [31], while the Iowa Women's Health Study with 34,492 postmenopausal women showed high whole-grain intake was associated with reduced risk of CHD related death [28,32]. The ARIC study, a prospective study with 15,792 individuals aged 45−64 years showed an inverse relationship with whole-grain intake and risk of mortality and the incidence of coronary artery disease [30]. In the Health Professionals Follow-Up Study with 43,757 men for every 10 g increase in cereal fiber intake there was a 29% reduced risk of heart attack [33]. A meta-analysis by Mellen and colleagues [34] evaluated outcomes from seven prospective cohorts and relationship with clinical cardiovascular events. Greater whole-grain intake (pooled average 2.5 servings/d vs 0.2 servings/d) was associated with a 21% lower risk of CVD events [Odds Ratio = 0.79 (95% CI: 0.73−0.85)] after adjusting for cardiovascular risk factors.

7.1.3 Biological Mechanisms of Benefit of Whole Grains

The association between whole-grain cereal consumption and reduced risk of CVD is not fully understood, but is likely mediated through a synergy of multiple pathways. This could be through a direct effect on improvements in CVD risk factors like glycemic control, blood pressure, dyslipidemia, and weight control [35]. A meta-analysis which pooled results from 21 randomized controlled trials found that a higher whole-grain intake lowers fasting blood glucose, insulin, total and LDL cholesterol, blood pressure, and weight gain.

More specifically, whole grains may impact chronic disease risk through the action of dietary fiber [36] and individual whole-grain components. Whole grains contain resistant starch, lignans, phytoestrogens, vitamins, minerals, trace elements, polyphenols and carotenoids, all which are cardioprotective and are likely to play a role in blood pressure lowering and possibly modulation of blood lipid levels [22,35,37].

The low-energy density and lower glycemic index of whole grains compared with refined grains may also increase satiety and prevent overeating to assist with weight control. Increasing postmeal satiety and reducing hunger has been shown to result in a lower energy intake. In 10 obese and 10 nonobese individuals with low-energy-density diets high in whole-grain foods or high in energy density showed that in the low-density diet, eating time was longer and satiation reached at a lower energy intake [38,39]. Satiety may also be influenced by carbohydrate oxidation. A diet high in cereal and legume intake resulted in a lower and delayed rise of postprandial carbohydrate oxidation and also a reduced feeling of hunger compared with a diet of refined carbohydrate [40].

The soluble fiber content of whole grains dampens the postprandial glycemic response by slowing the digestion of carbohydrate that may play a role in reducing the risk of type 2 diabetes mellitus. Similarly, the lower glycemic index of a food causes a lower and steady rise in postprandial blood glucose levels and reduces insulin release, giving greater glycemic control [41]. Moreover, the high fiber content of whole-grain foods may have a subsequent effect on lower day-long glucose concentrations, postprandial glycemic response, circulating insulin concentrations, and improved insulin sensitivity, perhaps related to the gel-forming properties of soluble fiber and subsequent impact on carbohydrate absorption.

7.1.4   Conclusions

Where previous research has provided limited and somewhat inconclusive evidence surrounding the benefits of whole-grain cereal consumption, there is a growing evidence base from epidemiological data supporting the consumption of whole grains for reduced risk of CVD and CVD-related mortality. Dietary patterns rich in whole-grain cereals such as the Mediterranean Diet reduce cardiovascular events, risk of type 2 diabetes mellitus and improve CVD risk factors through mechanisms related to satiety, glycemic control, antiinflammatory, and antioxidant pathways.

7.1.5   Cereal Consumption Recommendations

Recommendations for the consumption of whole-grain cereals should be widely supported by governing bodies, policy makers, food industry, and healthcare professionals.

The current recommendations of cereals in the Mediterranean Diet are of one to two serves of breads, rice, pasta, couscous, and other whole-grain cereals at every meal. If you do not consume whole-grain cereals regularly, begin little by little. Incrementing fiber content may cause some uncomfortableness at first, until the body adjusts to the new situation. You can add more whole grains to your diet following these tips:

- Substitute white bread for whole-grain bread. Make sure it is in fact whole grain, since some breads made with refined white flour and some bran added claim to be whole grain. If you are not used to to the flavor, you can alternate white and whole grains, benefitting equally from their consumption.
- If your breakfast consists primarily of cereal, try incorporating oats and other whole-grain cereals in the morning.
- Include some day of the week whole-grain rice and pasta. Whole-grain cereals can also be incorporated as a side dish to other food, instead of French fries, for example.
- If you usually make baked goods at home, cakes or muffins, use whole-grain flour instead of refined flour. Also, when making homemade pizza, use whole-grain flour for the crust.

- A healthy alternative for a snack could be whole-grain bread sticks seasoned with herbs and sesame seeds, instead of other snacks and baked goods.
- If there are children in the house, you can also offer them whole-grain cereals, such as the bread at lunch or as a snack or plates of rice and pasta, since these foods are more often accepted.
- Careful with food labels: not all whole-grain products are healthy, some products may not only contain whole grains, but also include unhealthy fats in their composition. Always read food labels to assure whole-grain flour is the primary ingredient.

7.2 LEGUMES AND CARDIOVASCULAR DISEASE

7.2.1 Overview and Introduction of Legumes in the Mediterranean Diet

There is abundant scientific evidence in the recent years that indicates that legumes optimizes the protective effect against cardiovascular risk. The health benefits arise from their characteristics, such as their low saturated fat content and high content of essential nutrients and phytochemicals, as well as to displacement effects when they are substituted for animal products in the diet [42]. The beneficial effects in blood lipid, glucose levels, insulin resistance, and in inflammatory and oxidative processes exhibited by the legumes bioactive compounds such as, fibers, phenolic compounds, tocopherols, carotenoids, some minerals has stimulated research on the mechanisms of action of these substances through distinct animal and human studies.

The legume (or pulse) family (*Leguminosae*) consists of plants that produce a pod with seeds inside. In this chapter, the term "legumes" is used to describe the seeds of these plants. Common edible legumes include dry beans, broad beans, dry peas, chickpeas, lentils, soybeans, lupins, mung beans, lotus, sprouts, alfalfa, green beans, peas, and peanuts.

The terms "legumes" and "pulses" are used interchangeably because all pulses are considered legumes but not all legumes are considered pulses. The term "pulse," as described in the Food and Agriculture Organization definition, is exclusively for crops harvested solely for the dry seed of leguminous plants. This excludes legumes used for oil extraction (soybean, peanut) and those harvested green for food (green beans, peas, and sprouts); the latter being classified as vegetables [43].

Legumes are uniquely low in fat and rich in both protein and complex carbohydrates including dietary fibers displaying a low glycemic index [44,45]. Also they are a good source of many essential nutrients, including vitamins, minerals, antioxidants, and other bioactive compounds. This food group contains a number of bioactive substances, including enzyme inhibitors, lectins, phytates, oligosaccharides, and phenolic compounds that play metabolic roles in humans or animals consuming frequently these foods [46].

The consumption and production of legumes extends worldwide, including the Mediterranean area where the most widely consumed legumes are bean (*Phaseolus vulgaris*), lentil (*Lens culinaris*), peas (*Pisum sativum*), chickpea (*Cicer arietinum*), and faba bean (*Vicia faba*, known as broad bean or fava bean).

Among European countries, higher legume consumption is observed around the Mediterranean, with per capita daily consumption between 8 and 23 g, while in Northern Europe, the daily consumption is less than 5 g per capita. In Europe, legume consumption has increased in the last decade (annually averaging 3.9 kg per capita), with differences between countries. Greece, Portugal, and Spain (annually averaging 6 kg per capita) consume the most of legumes.

7.2.2   Legumes and Cardiovascular Disease: Epidemiological Evidence

It has been confirmed that the consumption of grains, including pulses, legumes, and legume-based diets, contribute to a balanced diet and can prevent some chronic diseases, including type 2 diabetes mellitus and CVD. There is growing evidence both in intervention trials and observational studies of legumes consumption and CVD incidence [47,48]. In fact, the PREDIMED trial demonstrated that consuming a Mediterranean diet that included at least three serves of legumes a week reduced the risk in the incidence of developing an initial major CVD event by 30% [4].

Also, several observational studies indicate subjects that eat legumes are less likely to develop heart disease. In this way, results from the National Health and Nutrition Examination Survey Epidemiologic Follow-up Study indicate that men and women who reported consuming legumes four or more times per week had a 22% and 11% lower risk of coronary heart disease and CVD, respectively, compared with those who consumed legumes less than once a week [49].

The Japan Collaborative Cohort Study followed over 60,000 adults for 13 years and found the highest bean intake (4.5 serves a week) was associated with a 16% reduction in total CVD risk and a 10% reduction in total mortality [50].

A Costa Rican study observed that myocardial infarction survivors and their matched controls had a 38% lower risk of myocardial infarction if they consume at least 1/3 cup of cooked beans per day. Consuming greater amounts than this provided the same risk reduction but did not provide additional protection against myocardial infarction [51]. In a five-country longitudinal study of food habits among the elderly, legume consumption was the best predictor of survival. Risk of death was reduced by 8% for each daily 20 g intake of legumes [52].

Several intervention studies have demonstrated also that legumes can reduce CVD risk factors such as cholesterol, blood pressure, inflammation, blood sugars, and weight. For example, frequent legume consumption is likely to have a significant beneficial effect on coronary heart disease risk, by reducing serum total cholesterol, LDL cholesterol, and triglyceride levels, and rising HDL cholesterol. A 2011 meta-analysis of clinical trials investigating nonsoy legumes reported significant decreases in total cholesterol, LDL-cholesterol, and triglycerides (mainly in men with high cholesterol). The 10 trials studied the effects of 80−440 g/day (1/2 to 2 cups) of chickpeas, pinto beans, baked beans, navy beans as well as flour from ground beans. All studies reported net decreases in total cholesterol with a mean reduction of 5.5% in total cholesterol and 6.6% in LDL cholesterol [53].

Legumes also appear to beneficially affect other CVD risk factors. A 2014 systematic review reported that two serves of legumes (~ 162 g/day) significantly lowered blood pressure in people with and without hypertension [54]. Nonsoy legume intake has been found to significantly lower CRP (inflammation marker) concentrations [55,56] and one large scale cohort study reported an association between soy isoflavones and reduced CRP [57,58].

Recently, data have shown that legume consumption is inversely associated with serum concentrations of adhesion molecules and inflammatory biomarkers (serum concentrations of high-sensitive C reactive protein, tumor necrosis factor alpha, and interleukin-6 among Iranian women) [59]. The favorable association of legume consumption with molecule adhesion and inflammatory biomarkers might be explained by the low glycemic index values of legumes. The biologic mechanisms underlying these relations are not entirely understood. One possible mechanism through which dietary factors have been reported to influence metabolic health is a modulation of inflammation and endothelial function [60].

7.2.3 Nutritional Composition, Dietary Fibers, and Phytochemicals of Legumes: Their Role on Cardiovascular Risk

The physiological effects of different legumes vary significantly. These differences may result from the polysaccharides composition, in particular, the quantity and variety of dietary fibers and starch, protein content, and variability in phytochemical compounds [46]. The majority of legumes contain phytochemicals: bioactive compounds, including enzyme inhibitors, phytohemagglutinins (lectins), phytoestrogens, oligosaccharides, saponins, and phenolic compounds, which play metabolic roles in humans who frequently consume these foods. Dietary intake of phytochemicals may provide health benefits, protecting against numerous diseases or disorders, such as coronary heart disease, type 2 diabetes, high blood pressure, and inflammation. The synergistic or antagonistic effects of these phytochemical mixtures from food legumes, their interaction with other components of the diet, and the mechanism of their action have remained a challenge with regard to understanding the role of legumes in health and disease.

Nutrient content of legumes is exposed in Tables 7.1 and 7.2.

7.2.3.1   Legume Proteins

Legume seeds, compared to cereal grains, are rich in high-quality protein, providing people with a highly nutritious food resource. Dietary proteins also play a bioactive role by themselves and/or can be the precursors of biologically active peptides with various physiological functions [44]. The protein content of legume seeds ranges from 17% to 20% (dry weight) in peas and beans (6−8% when soaked and boiled) up to 38%−40% in soybean, depending on the species, similar to that of meats (18%−25%).

Legume seeds contain several comparatively minor proteins, including protease and amylase inhibitors, lectins, lipoxygenase, defense proteins, and others, which for various reasons are relevant to the nutritional/functional quality of the seeds. Enzyme inhibitors can diminish protein digestibility, and lectins can reduce nutrient absorption, but both have little effect after cooking. Some phenolic compounds can reduce protein digestibility and mineral bioavailability [44].

TABLE 7.1 Approximate Composition of Macronutrient of Various Legumes

Per 100 g of Boiled Legumes	Total Proteins (g)	Carbohydrate (g)	Total Fibers (g)	Total Lipids (g)	Resistant Starch (g)
Black beans	7.6	22.1	7.5	0.54	1.7
Chickpeas	6.5	21.9	6.2	2.5	2.6
Peas	6.8	18.2	2.0	0.51	-
Kidney beans	7.7	21.5	5.7	0.58	2.0
Lentils	8.9	18.9	7.8	0.38	3.4
Broadbeans	7.3	19.6	6.6	0.5	-
Navy beans	7.5	25.0	5.9	0.6	4.2
Pinto beans	7.7	18.3	7.7	0.51	1.9

Source: Adapted from Spanish Food Composition tables.

TABLE 7.2 Approximate Nutrient Content of Selected Beans

Per 100 g Boiled Legumes	Iron (mg)	Zinc (mg)	Calcium (mg)	Potassium (mg)	Magnesium (mg)	Folate (mg)
Black beans	1.8	0.96	23	305	60	128
Chickpeas	2.4	1.2	40	239	39	141
Peas	1.6	0.6	26	268	41	59
Kidney beans	2.0	0.9	31	358	37	115
Lentils	3.3	1.3	19	365	35	179
Broad beans	2.3	0.9	16	478	40	78
Navy beans	2.15	0.9	63	354	48	127
Pinto beans	1.8	0.8	40	373	43	147

Source: Adapted from Spanish Food Composition tables.

7.2.3.2 Legume Fibers

Legumes are a very good source of dietary fibers. Dietary fibers include resistant starch, nonstarch polysaccharides (cellulose, hemicellulose, pectin, gums, and b-glucans), nondigestible oligosaccharides, and lignin [61]. The ratio of soluble to insoluble fibers in legumes is comparable to that of grains (approximately 1:3 for both). High consumption of soluble fibers is associated with a decrease in serum total cholesterol, in LDL cholesterol, and is inversely correlated with coronary heart disease mortality rates [10,23,62]. In addition, an important consumption of dietary fibers, in particular resistant starch, is related with improved glucose tolerance and insulin sensitivity as we have already mentioned throughout the chapter [63]. It has been suggested that a state of satiety may be reached faster and last longer after ingestion of high-fiber foods, because they are bulkier and take longer to eat than lower fiber foods and delay gastric emptying.

Legumes contain a considerable amount of resistant starch, which is any starch that resists to digestion by amylase in the small intestine and progresses to the large intestine for fermentation by the gut bacteria [64,65]. The resistant starch content of beans is much higher than in commonly consumed grains, most likely because of their high ratio of amylose to amylopectin; amylose is a nonbranched, linear polymer of glucose units that is less readily digested than amylopectin. The resistant starch content from legumes varies between 1.7 g/100 g of boiled weight in black beans and 4 g in navy beans and chickpeas and lentils with 2.6 and 3.4 g/100 g of boiled legumes [64,65].

In addition to a high resistant starch content, legumes also have a higher ratio of slow-digestible to rapid-digestible starch, compared to other carbohydrate foods [65]. Legumes generally have a low glycemic index compared with other carbohydrate-rich foods, likely a result of both their resistant starch and fiber content [63]. The glycemic index of legumes ranges from 29 to 38 compared with 50 for brown rice and 55 for rolled oats [63].

The low glycemic index of legumes can potentially produce clinically relevant benefits related to CVD. Resistant starch is associated with reduced glycemic response, which can be beneficial to insulin-resistant individuals and those with diabetes [66]. In an intervention study participants with diabetes were instructed to increase their legume intake by at least 1 cup/d, glycated hemoglobin (Hb A1c) values decreased by 0.5% compared with a decrease of 0.3% in response to supplementation with wheat fiber [67]. Changes in Hb A1c concentrations of as little as 1% were associated with as much as a 15%−18% reduced risk of ischemic heart disease in people with type 2 diabetes [68].

7.2.3.3 Phytochemicals in Legumes

Legumes contain, in addition to the health-promoting components (fibers, proteins, resistant starch, and minerals), numerous phytochemicals endowed with useful biological activities. Phytochemicals in legumes includes polyphenols, flavonoids, isoflavonoids, anthocyanidins, saponins, phytoestrogens, terpenoids, carotenoids, limonoids, phytosterols, glucosinolates, phytohemagglutinin, and phytic acid [46]. Some data have reported that phytochemicals in legumes may have a beneficial effect on cardiovascular risk factors (Table 7.3).

The consumption of pulse grains has been reported to lower serum cholesterol and increase the cholesterol saturation levels in the bile [83]. A dietary study conducted on humans over a 7-week period showed that serum LDL cholesterol was significantly reduced during the consumption of a diet consisting of beans, lentils, and field peas [69]. The study showed that consumption of pulses lowered LDL cholesterol by partially interrupting the enterohepatic circulation of the bile acids and enhancing the cholesterol saturation by increasing the cholesterol hepatic secretion. Fenugreek (*Trigonella foenum graecum*) and isolated fenugreek fractions have been shown to act as hypocholesterolaemic agents in both animal and human studies as well as Faba beans (*V. faba*) also have lipid-lowering effects [84]. It has been suggested that these novel sources of legumes may provide health benefits when included in the daily diet [84].

TABLE 7.3 Phytochemicals of Legumes and Their Effects on Cardiovascular Risks

Antihyperlipidemic Effects

Phytochemicals	Reported Effects	References
Phytosterols: diosgenin; campesterol; brassicasterol; sitosterol; stigmasterol; uggulsterone	• Reduce blood cholesterol levels: total cholesterol and LDL cholesterol • Inhibit cholesterol absorption • Raise HDL cholesterol	Duane [69] Sanclemente [70] Sanclemente [71] Prakash [72]
Saponins; phytic acid	May reduce cholesterol through the formation of an insoluble complex with cholesterol and thus preventing absorption in the intestine	Prakash [72] Shimelis [73] Shi [74]

Antioxidant Effects

Phenolic compounds	• Free radical scavenging activity • Inhibit autoxidation of unsaturated lipid, thus preventing the formation of oxidized LDL • Modify LDL oxidation	Marathe [75] Amic [76] Prakash [72] Scalbert [77]
Flavonoids: isoflavonoids (soybeans)	Strong natural antioxidants with free radical scavenging activity; exert synergistic actions in scavenging free radicals	Prakash [72]

Weight and Glucose Management

Phytic acid	May have health benefits for diabetes patients by decreasing blood glucose response by: • reducing the rate of starch digestion • slowing the gastric emptying	Ganiyu [78] Roy [79] Lopez [80]
Phenolic compounds	May suppress growth of the adipose tissue through: • antiangiogenic activity • modulating adipocyte metabolism	Randhir [81] Mulvihill [82]
Phytochemicals (in general)	Reduce adipose tissue mass through: • Suppressing growth of adipose tissue • Inhibiting differentiation of preadipocytes • Stimulating lipolysis • Inducing apoptosis of existing adipocytes	Marathe [75]

7.2.3.3.1 Phytic Acid

Phytic acid exhibits antioxidant activity and protects against DNA damage; phenolic compounds have antioxidant and other important physiological and biological properties, and galactooligosaccharides may elicit prebiotic activity [78]. These compounds can have complementary and overlapping mechanisms of action, including regulation of detoxifying enzymes, stimulation of the immune system functionality, regulation of lipid metabolism, antioxidant, antimutagen, and antiangiogenic properties [78,84]. Although lectins and protease inhibitors are traditionally considered as protein antinutritional compounds, data have shown their potential in the treatment and/or the prevention of obesity and hypertension, so the use of the term "antinutritional" has been reconsidered [75,79].

7.2.3.3.2 Phenolic Compounds

Legumes contain a number of polyphenolic compounds (tannins, phenolic acids, and flavonoids) that may confer a variety of health benefits. As noted, many polyphenols are potent antioxidants, as showed in a previous analysis of 25 different types of beans, total antioxidant activity correlated directly with their polyphenol content [75]. As showed further numerous studies reveal that selected polyphenols exhibit strong protective actions on many pathological conditions, particularly those triggered by oxidative stress, such as CVD and metabolic disorders. The major sources of dietary polyphenols are cereals, legumes, oilseeds, fruits, vegetables, and some beverages.

In a study where Mediterranean legumes were analyzed [46], the simple polyphenols detected in various legumes are mainly phenolic acids and flavonoids [46]. The concentrations of simple polyphenols determined in these cooked legumes ranged from 321 to 2404 mg/100 g fresh weight in green split peas and big lentils, respectively, and decreased in the following order: lentils > chickpeas > pinto beans > white beans varieties, broad beans.

Legumes contain varied amounts of polyphenols and possess a wide range of antioxidant activity. Cowpea (brown and red), horse gram, common beans (black, red, brown, and beige), soybean, and fenugreek show excellent antioxidant activity. While faba beans (*V. faba*) and mung beans (*P. aureus, V. radiatus*) may also be a good source of antioxidants, chickpea (cream, green, and big brown), pea (white and green), and lablabbean (cream and white) show very weak antioxidant potential [84]. Thus, most of the varieties having a light-colored seed coat, except soybean, exhibit low antioxidant activity, while legumes having dark-colored seed coats, do not always possess high antioxidant activity (e.g., moth bean, black pea, black gram, lentils). Natural polyphenols exert their beneficial health effects by their antioxidant activity, these compounds are capable to remove free radicals, chelate metal catalysts, activate antioxidant enzymes, reduce α-tocopherol radicals and inhibit oxidases [76]. Phenolic phytochemicals inhibit oxidation of unsaturated lipids, thus preventing the formed oxidized LDL, which is considered to induce CVD. The antiradical activity of phenolic compounds is principally based on the redox properties of their hydroxyl groups and the structural relationships between different parts of their chemical structure [85].

7.2.3.3.3 Flavonoids

Flavonoids have a similar structure to oestrogens and have the capacity to exert both estrogenic and antioestrogenic effects and provide possible protection against heart disease. The precursors of these substances are widespread in the plant kingdom, mainly found in *leguminosae* and are especially abundant in soybean and its products, legumes, whole grains, and cereals. Among the biological activities of the flavonoids, they act against free radicals, free radical-mediated cellular signaling, inflammation, and platelet aggregation.

7.2.3.3.4 Phytosterols

Phytosterols are natural compounds structurally similar to mammalian cell-derived cholesterol. The best dietary sources of phytosterols are unrefined vegetable oils, seeds, cereals, nuts, and legumes. Kalogeropoulos et al. [46] reported that phytosterols content of Mediterranean cooked dry legumes exhibit a different phytosterols profile, ranging from 13.5 mg/100 g fresh weight in black-eyed beans to 53.6 mg/100 g in lupins. β-Sitosterol predominates in all cases, comprising 50%−85% of determined phytosterols and present in high concentration in chickpeas (38.52 mg/100 g). Campesterol and D5-avenasterol concentrations represent, respectively, 11.9 and 5.89 mg/100 g in white lupins, while stigmasterol content is 7.97 mg/100 g in medium white beans.

A high intake of these compounds can also protect against atherosclerosis [70] and decrease serum total cholesterol and LDL cholesterol levels. Mechanistically, phytosterols compete with cholesterol for micelle formation in intestinal lumen and inhibit cholesterol absorption. Their influence on intestinal genes and transcription factors makes phytosterols key regulators in metabolism and cholesterol transport in the expression of liver genes [71].

7.2.3.3.5 Saponins

Saponins have been found in many edible legumes (lupins, lentils, and chickpeas, as well as soy, various beans, and peas) [86]. Saponins in food legumes, especially in beans, have varying degrees of hemolytic and foam-producing activity. The hyperlipidemic or hypolipidemic action of saponins has not been well-studied and the results can be conflicting. However, some studies suggest that saponins may reduce cholesterol through the formation of an insoluble complex with cholesterol, thus preventing its absorption in the intestine. Additionally, some saponins increase the excretion of bile acids, an indirect method in decreasing cholesterol or are hydrolyzed by intestinal bacteria to diosgenin, which may exert a beneficial effect [86].

7.2.3.3.6   Legume Phytochemicals and Other Cardiovascular Risk Factors

Some few studies suggest that the regular consumption of legumes reduces the oxidative damage based on results of studies on screening of in vitro antioxidant activity of legumes [75,76,87].

Also, it has been reported that legumes antioxidants provide beneficial cardiovascular protection by reducing platelet aggregation and blood clotting, acting as antiinflammatory agents and improving vascular endothelial function [48,87]. Some studies have examined whether the consumption of legumes containing diets have an effect on glucose and lipid metabolism or on hormones controlling their metabolism [81]. Indeed, mung beans (*P. aureus, V. radiatus*) and soy are thought to be beneficial as an antidiabetic low glycemic index food Fenugreek (*T. foenum graecum*) and isolated fenugreek fractions have been shown to act as hypoglycaemic (and hypocholesterolaemic) agents, in both animal and human studies [81,84].

The role of legume phytochemicals in weight loss has been also assessed. Data have shown that weight loss was achievable with energy-controlled diets that were high in cereals and pulses [88]. Their potential role in weight management is unclear, although studies indicate that certain phenolics interfere with enterocyte glucose absorption through interference with the glucose transporters. Also, dietary phytochemicals might be employed as antiobesity agents because they may suppress growth of adipose tissue, inhibit differentiation of preadipocytes, stimulate lipolysis, and induce apoptosis of existing adipocytes, thereby reducing adipose tissue mass [75]. Moreover, dietary polyphenols may suppress growth of adipose tissue through their antiangiogenic activity and by modulating adipocyte metabolism. Additionally, legumes contain appreciable amounts of squalene, α- and $\beta + \gamma$ tocopherol, and generally their fatty acids profile is favorable for a cardioprotective perspective [82].

These results indicate the possible beneficial effects on excess body weight, hyperinsulinemia, and hyperlipidemia, which are the major cardiovascular risk factors.

7.3   CONCLUSIONS

Legumes, as foods with potential beneficial compounds, are gaining a great interest on chronic disease prevention and management. Legumes are a rich source of proteins, dietary fibers, including resistant starch, micronutrients, and bioactive compounds. These health benefits are known to be associated with not only with fibers but also phytochemicals like polyphenols present in legumes and other bioactive compounds (phytosterols, saponins, etc.). The regular consumption of these foods is essential into chronic disease prevention and management, including CVD.

This review was focused on reporting evidence data about legumes consumption CVD both with epidemiological data and focusing also on the nutritional quality of legumes and the beneficial effects of their phytochemicals on hyperinsulinemia, hyperlipidemia, inflammation, and the oxidative stress, which are the major cardiovascular risk factors, commonly associated with obesity and type 2 diabetes.

It can be concluded that legumes intake as an important component of Mediterranean Diet may contribute greatly in the management and/or prevention of cardiometabolic risk. Efforts should be made to encourage legume consumption not only in the Mediterranean area but worldwide as mentioned by the Food and Agriculture Organization of the United Nations (FAO) which declared for 2016, the International Year of Pulses.

7.4 LEGUME CONSUMPTION RECOMMENDATIONS

Legumes have been basic foods within the traditional Mediterranean diet, and for its health benefits it is important to include it in the diet at least twice a week. A serving of dry legumes (chickpeas, lentils, beans, etc.) is equivalent to 60−80 g.

You can follow these recommendations to increase your legume intake:

- Include in your diet different types of legumes; there is a great variety: lentils, chickpeas, beans, peas, soy beans, and lima beans.
- Legumes are a very nutritious dish on their own, but if you cook them with vegetables and whole grains you will create an even richer nutrient dense dish. Avoid incorporating meats with fat and sausages, which have cholesterol and saturated fat.
- Dishes with legumes and fish nutritionally complement each other and have an important health interest for. This combination combines the health benefits of the cardiovascular healthy fats of the fish with the health benefits legumes offer.
- You can use legumes in purees and soups. Purees tend to give less problems of flatulence than whole legumes.
- When legumes are less appetizing as a hot plate, bean salads are refreshing and very nutritious. Combine them with a variety of vegetables and canned fish to create a nutritiously complete and healthy dish. They can also be used as an ingredient in cold soups.
- As a side dish to meats and fish, they provide a good source of fiber and plant protein to the dish.
- When there is not enough time to cook legumes, canned legumes is a good resource.

- Hummus and falafels are traditional Arabic dishes made with chickpeas. They can serve as a very healthy appetizer instead of other less recommended appetizers, such as pâtés and sausages.

REFERENCES

[1] Mediterranean Diet Pyramid: a lifestyle for today. Guidelines for adult population. <http://dietamediterranea.com/piramidedm/piramide_INGLES.pdf>; 2010.

[2] Bach-Faig A, Berry EM, Lairon D, Reguant J, Trichopoulou A, Dernini S, et al. Mediterranean Diet Foundation Expert Group. Mediterranean diet pyramid today. Science and cultural updates. Public Health Nutr 2011;14(12A):2274−84.

[3] Katcher HI, Legro RS, Kunselman AR, Gillies PJ, Demers LM, Bagshaw DM, et al. The effects of a whole grain-enriched hypocaloric diet on cardiovascular disease risk factors in men and women with metabolic syndrome. Am J Clin Nutr 2008;87:79−90.

[4] Estruch R, Ros E, Salas-Salvadó J, Covas MI, Corella D, Arós F, et al. PREDIMED Study Investigators. Primary prevention of cardiovascular disease with a Mediterranean diet. N Engl J Med 2013;368:1279−90.

[5] Itsiopoulos C, Brazionis L, Kaimakamis M, Cameron M, Best JD, O'Dea K, et al. Can the Mediterranean diet lower HbA1c in type 2 diabetes? Results from a randomized cross-over study. Nutr Metab Cardiovasc Dis 2011;21:740−7.

[6] Martínez-González MA, Salas-Salvadó J, Estruch R, Corella D, Fitó M, Ros E. PREDIMED INVESTIGATORS. Benefits of the Mediterranean Diet: insights from the PREDIMED study. Prog Cardiovasc Dis 2015;58:50−60.

[7] Salas-Salvadó J, Bulló M, Babio N, Martínez-González MÁ, Ibarrola-Jurado N, Basora J, et al. PREDIMED Study Investigators. Reduction in the incidence of type 2 diabetes with the Mediterranean diet: results of the PREDIMED-Reus nutrition intervention randomized trial. Diabetes Care 2011;34:14−19.

[8] Sofi F, Cesari F, Abbate R, Gensini GF, Casini A. Adherence to Mediterranean diet and health status: meta-analysis. BMJ 2008;337:a1344.

[9] Scarmeas N, Stern Y, Tang MX, Mayeux R, Luchsinger JA. Mediterranean diet and risk for Alzheimer's disease. Ann Neurol 2006;59:912−21.

[10] Buil-Cosiales P, Zazpe I, Toledo E, Corella D, Salas-Salvadó J, Diez-Espino J, et al. Fiber intake and all-cause mortality in the Prevencion con Dieta Mediterranea (PREDIMED) study. Am J Clin Nutr 2014;100:1498−507.

[11] Wu H, Flint AJ, Qi Q, van Dam RM, Sampson LA, Rimm EB, et al. Association between dietary whole grain intake and risk of mortality: two large prospective studies in US men and women. JAMA Intern Med 2015;175:373−84.

[12] McKeown NM, Meigs JB, Liu S, Wilson PW, Jacques PF. Whole-grain intake is favorably associated with metabolic risk factors for type 2 diabetes and cardiovascular disease in the Framingham Offspring Study. Am J Clin Nutr 2002;76:390−8.

[13] Koh-Banerjee P, Rimm EB. Whole grain consumption and weight gain: a review of the epidemiological evidence, potential mechanisms and opportunities for future research. Proc Nutr Soc 2003;62:25−9.

[14] Liese AD, Roach AK, Sparks KC, Marquart L, D'Agostino Jr RB, Mayer-Davis EJ. Whole-grain intake and insulin sensitivity: the Insulin Resistance Atherosclerosis Study. Am J Clin Nutr 2003;78:965−71.

[15] Jenkins DJ, Axelsen M, Kendall CW, Augustin LS, Vuksan V, Smith U. Dietary fibre, lente carbohydrates and the insulin-resistant diseases. Br J Nutr 2000;83(Suppl 1):S157−63.

[16] Ripsin CM, Keenan JM, Jacobs Jr DR, Elmer PJ, Welch RR, Van Horn L, et al. Oat products and lipid lowering. A meta-analysis. JAMA 1992;267:3317−25.

[17] Schultz K, Westereng B. Cholesterol-lowering benefits of a whole grain oat ready-to-eat cereal. Nutr Clin Care 1998;1:6−12.

[18] Appel LJ, Moore TJ, Obarzanek E, Vollmer WM, Svetkey LP, Sacks FM, et al. A clinical trial of the effects of dietary patterns on blood pressure. DASH Collaborative Research Group. N Engl J Med 1997;336:1117−24.

[19] Sacks FM, Svetkey LP, Vollmer WM, Appel LJ, Bray GA, Harsha D, et al. Effects on blood pressure of reduced dietary sodium and the Dietary Approaches to Stop Hypertension (DASH) diet. DASH-Sodium Collaborative Research Group. N Engl J Med 2001;344:3−10.

[20] Serra-Majem L, Bautista-Castano I. Relationship between bread and obesity. Br J Nutr 2015;113(Suppl 2):S29−35.

[21] Bautista-Castano I, Serra-Majem L. Relationship between bread consumption, body weight, and abdominal fat distribution: evidence from epidemiological studies. Nutr Rev 2012;70:218−33.

[22] Kelly SA, Summerbell CD, Brynes A, Whittaker V, Frost G. Wholegrain cereals for coronary heart disease. Cochrane Database Syst Rev 2007; CD005051.

[23] Brown L, Rosner B, Willett WW, Sacks FM. Cholesterol-lowering effects of dietary fiber: a meta-analysis. Am J Clin Nutr 1999;69:30−42.

[24] Belobrajdic DP, Bird AR. The potential role of phytochemicals in wholegrain cereals for the prevention of type-2 diabetes. Nutr J 2013;12:62.

[25] Liu S, Manson JE, Stampfer MJ, Hu FB, Giovannucci E, Colditz GA, et al. A prospective study of whole-grain intake and risk of type 2 diabetes mellitus in US women. Am J Public Health 2000;90:1409−15.

[26] Meyer KA, Kushi LH, Jacobs Jr DR, Slavin J, Sellers TA, Folsom AR. Carbohydrates, dietary fiber, and incident type 2 diabetes in older women. Am J Clin Nutr 2000;71:921−30.

[27] Fung TT, Hu FB, Pereira MA, Liu S, Stampfer MJ, Colditz GA, et al. Whole-grain intake and the risk of type 2 diabetes: a prospective study in men. Am J Clin Nutr 2002;76:535−40.

[28] Jacobs Jr DR, Meyer KA, Kushi LH, Folsom AR. Is whole grain intake associated with reduced total and cause-specific death rates in older women? The Iowa Women's Health Study. Am J Public Health 1999;89:322−9.

[29] Jacobs Jr DR, Andersen LF, Blomhoff R. Whole-grain consumption is associated with a reduced risk of noncardiovascular, noncancer death attributed to inflammatory diseases in the Iowa Women's Health Study. Am J Clin Nutr 2007;85:1606−14.

[30] Steffen LM, Jacobs Jr DR, Stevens J, Shahar E, Carithers T, Folsom AR. Associations of whole-grain, refined-grain, and fruit and vegetable consumption with risks of all-cause mortality and incident coronary artery disease and ischemic stroke: the Atherosclerosis Risk in Communities (ARIC) Study. Am J Clin Nutr 2003;78:383−90.

[31] Liu S, Stampfer MJ, Hu FB, Giovannucci E, Rimm E, Manson JE, et al. Whole-grain consumption and risk of coronary heart disease: results from the Nurses' Health Study. Am J Clin Nutr 1999;70:412−19.

[32] Jacobs Jr DR, Meyer KA, Kushi LH, Folsom AR. Whole-grain intake may reduce the risk of ischemic heart disease death in postmenopausal women: the Iowa Women's Health Study. Am J Clin Nutr 1998;68:248−57.

[33] Rimm EB, Ascherio A, Giovannucci E, Spiegelman D, Stampfer MJ, Willett WC. Vegetable, fruit, and cereal fiber intake and risk of coronary heart disease among men. JAMA 1996;275:447−51.

[34] Mellen PB, Walsh TF, Herrington DM. Whole grain intake and cardiovascular disease: a meta-analysis. Nutr Metab Cardiovasc Dis 2008;18:283−90.

[35] Seal CJ. Whole grains and CVD risk. Proc Nutr Soc 2006;65:24−34.

[36] Dahl WJ, Stewart ML. Position of the Academy of Nutrition and Dietetics: health implications of dietary fiber. J Acad Nutr Diet 2015;115:1861−70.

[37] Reicks M, Jonnalagadda S, Albertson AM, Joshi N. Total dietary fiber intakes in the US population are related to whole grain consumption: results from the National Health and Nutrition Examination Survey 2009 to 2010. Nutr Res 2014;34:226−34.

[38] Duncan KH, Bacon JA, Weinsier RL. The effects of high and low energy density diets on satiety, energy intake, and eating time of obese and nonobese subjects. Am J Clin Nutr 1983;37:763−7.

[39] Weinsier RL, Johnston MH, Doleys DM, Bacon JA. Dietary management of obesity: evaluation of the time-energy displacement diet in terms of its efficacy and nutritional adequacy for long-term weight control. Br J Nutr 1982;47:367−79.

[40] Sparti A, Milon H, Di Vetta V, Schneiter P, Tappy L, Jéquier E, et al. Effect of diets high or low in unavailable and slowly digestible carbohydrates on the pattern of 24-h substrate oxidation and feelings of hunger in humans. Am J Clin Nutr 2000;72:1461−8.

[41] Giacco R, Costabile G, Della Pepa G, Anniballi G, Griffo E, Mangione A, et al. A whole-grain cereal-based diet lowers postprandial plasma insulin and triglyceride levels in individuals with metabolic syndrome. Nutr Metab Cardiovasc Dis 2014;24:837−44.

[42] Messina V. Nutritional and health benefits of dried beans. Am J Clin Nutr 2014;100 (Suppl 1):437S−442SS.

[43] FAO. Definition and classification of commodities, 4 pulses and derived products. <http://www.fao.org/es/faodef/fdef04e.htm>; 2004.

[44] Duranti M. Grain legume proteins and nutraceutical properties. Fitoterapia 2006;77:67−82.

[45] Guéguen J, Cerletti P. Proteins of some legume seeds, soybean, pea, fababean and lupin. In: Hudson BJF, editor. New and developing sources of food proteins. New York: Chapman and Hall; 1994. p. 145−93.

[46] Kalogeropoulos N, Chiou A, Ioannou M, Karathanos VT, Hassapidou M, Nikolaos K, et al. Nutritional evaluation and bioactive microconstituents (phytosterols, tocopherols, polyphenols, triterpenic acids) in cooked dry legumes usually consumed in the Mediterranean countries. Food Chem 2010;121:682−90.

[47] Rizkalla SW, Bellisle F, Slama G. Health benefits of low glycaemic index foods, such as pulses, in diabetic patients and healthy individuals. Br J Nutr 2002;88(Suppl 3):S255−62.

[48] Flight I, Clifton P. Cereal grains and legumes in the prevention of coronary heart disease and stroke: a review of the literature. Eur J Clin Nutr 2006;60:1145−59.

[49] Bazzano LHJ, Ogden LG, Loria C, Vupputuri S, Myers L, Whelton PK. Legume consumption and risk of coronary heart disease in U.S. men and women: NHANES I Epidemiologic Follow-up Study. Arch Inter Med 2001;161:2573−8.

[50] Nagura J, Iso H, Watanabe Y, Maruyama K, Date C, Totoyoshima H, et al. Fruit, vegetable and bean intake and mortality from cardiovascular disease among Japanese men and women: the JACC Study. Br J Nutr 2009;102:285−92.

[51] Monge-Rojas R, Mattei J, Fuster T, Willett W, Campos H. Influence of sensory and cultural perceptions of white rice, brown rice and beans by Costa Rican adults in their dietary choices. Appetite 2014;81:200−8.

[52] Kabagambe EK, Baylin A, Ruiz-Narvarez E, Siles X, Campos H. Decreased consumption of dried mature beans is positively associated with urbanization and nonfatal acute myocardial infarction. J Nutr 2005;135:70−5.

[53] Bazzano LA, Thompson AM, Tees MT, Nguyen CH, Winham DM. Non-soy legume consumption lowers cholesterol levels: a meta-analysis of randomized controlled trials. Nutr Metab Cardiovasc Dis 2011;21:94−103.

[54] Jayalath VH, de Souza RJ, Sievenpiper JL, Ha V, Chiavaroli L, Mirrahimi A, et al. Effect of dietary pulses on blood pressure: a systematic review and meta-analysis of controlled feeding trials. Am J Hypertens 2014;27:56−64.

[55] Beavers KM, Jonnalagadda SS, Messina MJ. Soy consumption, adhesion molecules, and pro-inflammatory cytokines: a brief review of the literature. Nutr Rev 2009;67:213−21.

[56] Salehi-Abargouei A, Saraf-Bank S, Bellissimo N, Azadbakht L. Effects of non-soy legume consumption on C-reactive protein: a systematic review and meta-analysis. Nutrition 2015;31:631−9.

[57] Nicastro H, Mondul A, Rohrmann S, Platz E. Associations between urinary soy isoflavonoids and two inflammatory markers in adults in the United States in 2005-2008. Cancer Causes Control 2013;24:1185−96.

[58] Lee YP. Effects of lupin kernel flour-enriched bread on blood pressure: a controlled intervention study. Am J Clin Nutr 2009;89:766.

[59] Esmaillzadeh A, Azadbakht L. Legume consumption is inversely associated with serum concentrations of adhesion molecules and inflammatory biomarkers among Iranian women. J Nutr 2012;142:334−9.

[60] Giugliano D, Ceriello A, Esposito K. The effects of diet on inflammation emphasis on the metabolic syndrome. J Am Coll Cardiol 2006;48:677−85.

[61] Tharanathan RN, Mahadevamma S. Legumes—a boon to human nutrition. Trends Food Sci Tech 2003;14:507−18.

[62] National Cholesterol Education Program Expert Panel. Third report of the National Cholesterol Education Program (NCEP) Expert Panel on detection, evaluation, and treatment of high blood cholesterol in adults (Adult Treatment Panel III): final report. Circulation 2002;106:3140−1.

[63] Jenkins DJ, Wolever TM, Taylor RH, Barker HM, Fielden H. Exceptionally low blood glucose response to dried beans: comparison with other carbohydrate foods. BMJ 1980;281:578−80.

[64] Yadav BS, Sharma A, Yadav RB. Resistant starch content of conventionally boiled and pressure-cooked cereals, legumes and tubers. J Food Sci Technol 2010;47:84−8.

[65] Thorne MJ, Thompson LU, Jenkins DJ. Factors affecting starch digestibility and the glycemic response with special reference to legumes. Am J Clin Nutr 1983;38:481−8.

[66] Park OJ, Kang NE, Chang MJ, Kim WK. Resistant starch supplementation influences blood lipid concentrations and glucose control in overweight subjects. J Nutr Sci Vitaminol 2004;50:93−9.

[67] Jenkins DJ, Kendall CW, Augustin LS, Mitchell S, Sahye-Pudaruth S, Blanco Mejia S, et al. Effect of legumes as part of a low glycemic index diet on glycemic control and cardiovascular risk factors in type 2 diabetes mellitus: a randomized controlled trial. Arch Intern Med 2012;172:1653−60.

[68] Syed IA, Khan WA. Glycated haemoglobin—a marker and predictor of cardiovascular disease. J Pak Med Assoc 2011;61:690−5.

[69] Duane WC. Effects of legume consumption on serum cholesterol, biliary lipids, and sterol metabolism in humans. J Lipid Res 1997;38:1120−8.

[70] Sanclemente T, Marques-Lopes I, Fajó-Pascual M, Cofán M, Jarauta E, Ros E, et al. Naturally-occurring phytosterols in the usual diet influence cholesterol metabolism in healthy subjects. Nutr Metab Cardiovasc Dis 2012;22:849−55.

[71] Sanclemente T, Marques-Lopes I, Fajó-Pascual M, Cofán M, Jarauta E, Ros E, et al. A moderate intake of phytosterols from habitual diet affects cholesterol metabolism. J Physiol Biochem 2009;65:397−404.

[72] Prakash D, Gupta C. Role of phytoestrogens as nutraceuticals in human health. Pharmacologyonline 2011;1:510−23.

[73] Shimelis A. Potential health benefits and problems associated with phytochemicals in food legumes. East Afr J Sci 2009;3:116−33.

[74] Shi J, Arunasalam K, Yeung D, Kakuda Y, Mittal G, Jiang Y. Saponins from edible legumes: chemistry, processing, and health benefits. J Med Food 2004;7:67−78.

[75] Marathe SA, Rajalakshmi V, Jamdar SN, Sharma A. Comparative study on antioxidant activity of different varieties of commonly consumed legumes in India. Food Chem Toxicol 2011;49:2005−12.

[76] Amic D, Davidovic-Amic D, Beslo D, Trinajstic N. Antioxidant and antimicrobial properties of Telfairia occidentalis (fluted pumpkin) leaf extracts. Croatia Chem Acta 2003;76:55−61.

[77] Scalbert A, Manach C, Morand C, Remesy C. Dietary polyphenols and the prevention of diseases. Crit Rev Food Sci Nutr 2005;45:287−306.

[78] Ganiyu O. Antioxidant properties of some commonly consumed and underutilized tropical legumes. Eur Food Res Technol 2006;224:61−5.

[79] Roy F, Boye JI, Simpson BK. Bioactive proteins and peptides in pulse crops: pea, chickpea and lentil. Food Res Intern 2010;43:432−532.

[80] Lopez G, Pedro M, Garzon de la Mora P, Wysocka W, Maiztegui B, Alzugaray ME, et al. Quinolizidine alkaloids isolated from Lupinus species enhance insulin secretion. Eur J Pharmacol 2004;504:139−42.

[81] Randhir R, Shetty K. Mung beans processed by solid-state bioconversion improves phenolic content and functionality relevant for diabetes and ulcer management. Innovat Food Sci Emerg Technol 2007;8:197−204.

[82] Mulvihill EE, Huff MW. Antiatherogenic properties of flavonoids: implications for cardiovascular health. Can J Cardiol 2010;26(Suppl A):17A−21A.

[83] Rochfort S, Panozzo J. Phytochemicals for health, the role of pulses. J Agric Food Chem 2007;55:7981−94.

[84] Zecharia M, Aliza HS. New legume sources as therapeutic agents. Br J Nutr 2002;88:287−92.

[85] Yao Y, Cheng X, Wang L, Wang S, Ren G. Biological potential of sixteen legumes in China. Int J Mol Sci 2011;12:7048−58.

[86] Fenwick DE, Oakenfull D. Saponin content of food plants and some prepared foods. J Sci Food Agric 1983;34:186−91.

[87] King DE, Egan BM, Woolson RF, Mainous III AG, Al-Solaiman Y, Jesri A. Effect of a high-fiber diet vs. a fiber-supplemented diet on C-reactive protein level. Arch Intern Med 2007;167:502−6.

[88] Venn BJ, Perry T, Green TJ, Skeaff CM, Aitken W, Moore NJ. The effect of increasing consumption of pulses and wholegrains in obese people: a randomized controlled trial. J Am Coll Nutr 2010;29:365−72.

Chapter 8

More Fish, Less Meat

Mary K. Downer and Ana Sánchez-Tainta

8.1 EARLY RESEARCH ON FISH INTAKE AND CARDIOVASCULAR DISEASE

In the 1970s, the first research supporting the idea that fish consumption may be related to decreased risk of cardiovascular disease originated from descriptive epidemiological data. Scientists first noted that coronary heart disease (CHD) incidence and mortality rates varied by up to 10-fold or more across geographical areas, suggesting that differences in environments or modifiable lifestyle factors could influence CHD incidence and progression. Studies of Japanese migrants showed that when people from Japan—an area with traditionally low rates of CHD—moved to the United States, CHD incidence and mortality rates among this population increased, becoming more similar to rates of their new home in the United States. This was an even stronger indication that CHD was largely due to modifiable risk factors [1]. Around the same time, researchers began to notice strikingly low rates of cardiovascular disease among the Greenland Eskimos, and suspected that this may be due in part to their unusually high fish consumption [2]. Similarly, low rates of cardiovascular disease coupled with high fish consumption were also noted in Japan, furthering the hypothesis that high fish intake may have cardiovascular benefits [3]. These intriguing findings prompted many other early epidemiological studies to investigate the role of fish intake on cardiovascular disease, and they produced similar findings [4–6]. During the last several decades, the scientific community has conducted extensive research in this area. Findings have consistently suggested that fish consumption has powerful benefits for cardiovascular disease risk reduction.

8.2 NUTRITIONAL COMPOSITION OF FISH IN RELATION TO CARDIOVASCULAR BENEFITS

After these first epidemiological studies emerged, the initial hypothesized mechanism was that the high omega-3 fatty acid content of fish inhibited

The Prevention of Cardiovascular Disease through The Mediterranean Diet.
DOI: http://dx.doi.org/10.1016/B978-0-12-811259-5.00008-1

platelet function, which in turn reduced thrombus formation in coronary arteries. Since then, many animal-experimental, observational, and clinical studies sought to identify the mechanism(s) explaining the apparent protective benefit of fish consumption on cardiovascular disease. Conclusive evidence points to two long-chain omega-3 polyunsaturated fatty acids (omega-3 PUFAs) abundant in fish, eicosapentaenoic acid (EPA) and docosahexaenoic acid (DHA), as the chief components responsible for the cardiovascular benefits of fish consumption [7]. EPA and DHA are primarily determined by dietary consumption, and fish are the primary dietary source of both of these important omega-3 PUFAs [8]. Although alpha-linolenic acid (ALA), a plant-based essential fatty acid, can be converted to EPA and DHA, only 0.2%−8% [9] is actually converted. Thus, plants rich in ALA are not a very weak source of EPA and DHA.

Both of these omega-3 PUFAs are characterized by long hydrocarbon backbones, multiple double bonds, and a first double bond on the n-3 position [8]. Unsaturated fatty acids, including EPA and DHA, are in the *cis* formation, meaning they have hydrogen atoms on the same side of the acyl chain, which causes a larger "kink" in the fatty acyl chain, thus preventing unsaturated fatty acids from packing together. This is why unsaturated fats are not solid at room temperature [10]. These unique features also allow EPA and DHA to possess many unique biological properties that contribute to cardiovascular benefits. It should also be noted that docosapentaenoic acid (DPA), an omega-3 PUFA metabolite of EPA, is also present in smaller amounts in fish. However, endogenous metabolism is the primary determinant of circulating DPA levels, and little is known about DPA's physiological effects or its role in cardiovascular disease [8]. Many fish are also abundant in high-quality proteins, vitamins A, B, D, or E, iron, zinc, selenium, and iodine, but less is known about the relationship between these nutrients in relation to cardiovascular disease.

8.3 NUTRITIONAL COMPOSITION OF MEAT IN RELATION TO CARDIOVASCULAR BENEFITS

The nutritional composition of red and processed meat—which is often consumed instead of fish—is much different than fish. The differences dietary fatty acid content is likely the most relevant in relation to cardiovascular disease risk. In contrast to fish, red and processed meats are high in saturated fat and some types (e.g., cattle) contains small amounts of trans fat, both of which have been shown to substantially increase cardiovascular disease risk [1]. In comparison to unsaturated fatty acids, saturated fatty acids (SFAs) do not have any double bonds—they are fully saturated with hydrogen molecules, and are solid at room temperature. Trans fatty acids, like unsaturated fatty acids including EPA and DHA, also have double bonds but in the trans conformation, with hydrogen atoms on opposite sides of the acyl chain. A *trans* double bond causes trans fatty acids to be more similar in shape to SFA, with smaller "kinks" in the acyl chain to allow multiple trans fatty acids to pack together; this is why trans fats are solid at room temperature [10]. Typically, meat also tends to be more calorie dense and have higher sodium content compared to fish.

Two prevalent subtypes of meat implicated in increased cardiovascular disease risk are red meat and processed meat. In comparison to red meat, processed meat is slightly more calorie dense with a higher percent energy from fat and lower percent energy from protein, has less iron, slightly less cholesterol, approximately four times more sodium, and 50% more nonsalt preservatives such as nitrates, nitrites, and nitrosamines. Saturated fat content of red and processed meat is similar [11].

8.4 INTERMEDIATE PHYSIOLOGICAL EFFECTS OF FISH INTAKE

Omega-3 PUFAs have numerous physiological effects, which may play an intermediary role in reducing risk of cardiovascular disease. These include, but are not limited to, lowering plasma triglycerides, reducing heart rate and blood pressure, improving myocardial filling and efficiency, decreasing inflammation, and antiarrhythmic effects (Fig. 8.1). The processes by which these benefits occur are detailed below.

8.4.1 Lowering Plasma Triglycerides

Extensive evidence has demonstrated that consuming omega-3 PUFAs lowers plasma triglycerides [12]. When omega-3 PUFAs are consumed, hepatic very low-density lipoprotein (VLDL; "harmful cholesterol") synthesis leads to several beneficial processes. Fewer carbohydrates are converted to fat, fatty acid beta-oxidation increases, and fewer nonesterified fatty acids are sent to the liver. These processes create a limited availability of fatty acids for triglyceride synthesis. Additionally, circulating omega-3 PUFAs increase hepatic enzyme activity for phospholipid synthesis while decreasing activity for triglyceride synthesis. At a population level, increased omega-3 PUFA consumption is linearly associated with decreased plasma triglycerides. However, effects vary at an individual level; increasing omega-3 PUFAs seem to lower triglyceride levels more among those with already high omega-3 PUFA consumption. Over time, the triglyceride-lowering effects of sustained high omega-3 PUFA consumption may modestly decrease cardiovascular disease risk. However, at regular omega-3 PUFA intake levels, this impact is minor and an unlikely explanation for reduced cardiovascular disease outcomes [8].

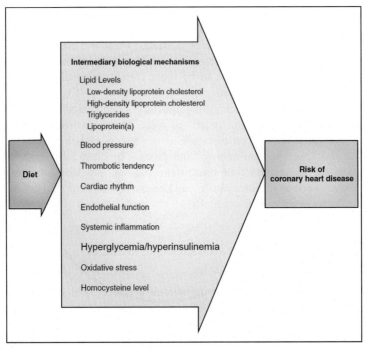

FIGURE 8.1 Potential pathways mediating the effects of diet on risk of coronary heart disease. *From Willett WC. Dietary fats and coronary heart disease. J Intern Med 2012;272(1):13−24.*

8.4.2 Reducing Heart Rate and Blood Pressure

Experimental studies have found that consuming omega-3 PUFAs directly benefits cardiac electrophysiological pathways, thereby reducing heart rate and blood pressure. Omega-3 PUFAs have also been shown to improve left ventricular diastolic filling, augment vagal tone, increase nitric oxide production, decrease vasoconstrictive responses to norepinephrine and angiotensin II, and improve vasodilatory responses and atrial compliance, all of which likely contribute to reduced heart rate and blood pressure [8].

8.4.3 Improving Myocardial Filling and Efficiency

Most evidence indicating that omega-3 PUFA consumption improves myocardial filling and efficiency comes from animal studies, although there is increasing human-based evidence as well. Both short-term and long-term effects have been observed. It is hypothesized that this long-term benefits result from reduced systemic vascular resistance with omega-3 PUFA consumption. In addition to this structural benefit, it is hypothesized that functional or metabolic mechanisms are responsible for the observed short-term benefits (i.e., immediate improvement in early diastolic filling) [8].

8.4.4   Decreasing Inflammation

Similar to triglyceride-lowering effects, it is unclear whether typical omega-3 PUFA consumption levels decrease inflammation enough to have clinically meaningful cardiovascular benefits. Several studies indicate that omega-3 PUFAs reduce eicosanoids, a family of proteins important in inflammatory responses, although their effect on cytokines, many of which are critical in inflammatory reactions, is less clear. Fish oil supplementation appears to alleviate pain related to rheumatoid arthritis. EPA, DHA, and other omega-3 fatty acids are also precursors to resolvins, protectins, and other components that may resolve inflammation. However, biomarkers may be unable to detect other local antiinflammatory effects, and much is unknown about the antiinflammatory effects of omega-3 PUFAs [8].

8.4.5   Antiarrhythmia Effects

It is likely that omega-3 PUFA consumption's effects on membrane ion channels and cell-cell connexins have direct benefits on atrial and ventricular myocyte electrophysiology. However, this benefit has not been confirmed because a reliable biomarker has not been identified, and the mechanisms remain unclear [8].

8.4.6   Visceral Adiposity

Higher intakes of PUFA and monounsaturated fatty acids (MUFA) have been associated with lower visceral adiposity [13]; moderate fish consumption is also associated with lower visceral adiposity [14]. Mechanisms by which this may occur remain unclear.

There is extensive evidence that fish intake confers the above cardiovascular benefits. In addition, there is also less extensive evidence that fish intake may decrease thrombosis and insulin resistance and improve endothelial and autonomic function.

8.5   INTERMEDIATE PHYSIOLOGICAL EFFECTS OF MEAT INTAKE

The following physiological effects of increased meat intake are established mechanisms that increase risk of cardiovascular disease.

8.5.1   Increased LDL: HDL Serum Cholesterol Ratio

Saturated fat from red and processed meat increases both LDL (associated with cardiovascular harm) and HDL (associated with cardiovascular benefits) cholesterol [1]. Dietary trans fatty acids also increase LDL cholesterol, but without increasing HDL cholesterol [10]; it likely even reduces HDL [15]. Dietary substitution analyses show that substituting mono- and polyunsaturated fats for trans and saturated fats reduces LDL cholesterol without reducing HDL or increasing triglycerides [1]. LDL/HDL ratios strongly predict negative cardiovascular consequences, such as leading to atheromatous plaques, which reduce blood flow to the heart by narrowing coronary arteries. Reduced blood flow from consequences of increased LDL/HDL ratio can ultimately lead to myocardial infarction (MI) [1]. However, more research is needed to determine whether ruminant trans fatty acids from animals vs trans fat from industrial partial hydrogenation of vegetable oils has different effects [16].

8.5.2   Increased Triglycerides

Saturated [17] and trans [10] fatty acids, present in red and processed meat, have been reported to increase triglyceride levels. A high triglyceride level is an established intermediary biological mechanism that increases risk of CHD and other cardiovascular disease outcomes [1].

8.5.3   Insulin Resistance, Hyperglycemia, Hyperinsulinemia

Red and processed meats are high in dietary cholesterol. Studies show that increased dietary cholesterol is associated with abnormal glucose metabolism [18]. Furthermore, preservatives such as nitrates and byproducts, which are high in processed meat, have been shown to impair glucose tolerance and reduce insulin secretion [19,20]. These irregularities in glucose metabolism and insulin levels directly lead to increased risk of type 2 diabetes. Diabetes is associated with a twofold to threefold increased risk of heart failure and other cardiovascular disease outcomes [21−23].

8.5.4   Body Weight

A 2014 systematic review and meta-analysis of observational studies investigating the relationship between red and processed meat and obesity found that both were associated with increased risk of obesity, higher body mass index, and larger waist circumference [24]. Diets high in saturated fats are shown to be associated with increased risk of obesity and body fat accumulation [25,26], which is likely the mechanism by which red and processed fat increase body weight to unhealthy levels. Ruminant trans fat, present in red and processed meat, may also be associated with weight gain [27,28].

8.5.5   Worse Arterial Compliance and Vascular Stiffness

Increased sodium intake from red and processed meats has been shown to decrease arterial compliance and increase vascular stiffness [29]. Nitrates and byproducts, high in processed meat, also promote atherosclerosis and vascular dysfunction [30].

8.5.6   Hypertension

Many nutritional components of red and processed meat increase blood pressure. Extensive evidence has established that increased dietary sodium increases blood pressure [31−33], and thus low sodium diets are one of the most effective ways to reduce blood pressure in hypertensive individuals [31]. A prospective study of US nurses reported a positive, linear relationship between red meat intake—which generally has lower sodium content compared to processed meat—and risk of hypertension [34]; prospective studies with more ethnically diverse populations including both men and women found similar results [35,36], as did cross-sectional epidemiological studies [35,37].

8.6   EPIDEMIOLOGICAL EVIDENCE—FISH INTAKE AND CARDIOVASCULAR DISEASE

Evidence consistently shows that fish intake as associated with a reduced risk of several cardiovascular disease outcomes. The strongest evidence for the association between fish intake and reduced risk of cardiovascular disease is for CHD mortality. CHD mortality includes documented or suspected fatal MI and sudden cardiac death, which is defined as a "sudden pulseless condition of presumed cardiac etiology" [7].

After the early ecological studies observing that cardiovascular disease was remarkably low in populations with high fish consumption, many more sophisticated studies investigated this relationship. There have been more prospective observational studies and randomized controlled trials (RCTs) investigating the association cardiovascular disease and fish or omega-3 PUFA intake than any other dietary component [8]. Descriptive studies, controlled feeding studies, animal studies, and more also contribute to this overwhelming body of evidence.

Beginning in the 1980s, the first prospective observational studies demonstrated the association between fish consumption and reduced risk of CHD mortality [4,38−40]. Other prospective observational studies indirectly examining this association have repeatedly produced consistent findings. A pooled analysis investigating dietary fats and CHD risk found that replacing SFA with PUFA was associated with the strongest reduction in risk [41]. Another pooled analysis comparing vegetarians (no fish, no meat), pescatarians (fish, no meat), and regular meat eaters (no restrictions) found that pescatarians had the lowest risk of CHD mortality [40,42].

There have been many RCTs investigating dietary fat and CHD mortality. However, many were unable to achieve meaningful long-term dietary change or differences across intervention arms [22,43,44] which is likely necessary to modify CHD mortality outcomes. Others were not adequately powered to detect effects [44–49]. Nevertheless, the few successful RCTs have also found that diets high in MUFA/PUFA and low in saturated fat—which is true of fish—are associated with much lower rates of CHD mortality and other cardiovascular disease outcomes. A meta-analyses of RCTs investigating omega-3 fatty acids and CHD mortality concluded that there was a strong inverse association [8]. The Oslo Heart Study, an RCT of 16,202 men beginning in 1972 found that the dietary intervention group, which achieved a reduction in saturated fat and dietary cholesterol and a small increase in PUFA, had a much lower (47%) risk of both nonfatal MI and CHD mortality compared to the control group [50]. Between 1988 and 1992, the Lyon Heart Study randomized participants to a typical low-fat diet or Mediterranean Diet also observed a large (70%) reduced risk in CHD [51]. However, this intervention was multifaceted and thus it is difficult to determine which aspect of the diet was responsible for observed risk reduction. More recently, the PREDIMED trial successfully randomized over 7000 Spanish adults at high risk for cardiovascular disease to a low-fat diet or Mediterranean-type diet, which encouraged high consumption of fish, among other things. Similar to results from observational studies, PREDIMED also found that MUFA and PUFA intake were associated with lower cardiovascular disease and death, while saturated fat and trans fat had the opposite effect; replacing saturated fat with MUFA and PUFA was also associated with reduced risk [52], as expected.

There is evidence suggesting that fish intake is also associated with reduced risk of nonfatal cardiovascular disease outcomes. A summary of large primary and secondary dietary prevention studies estimates that a 10% reduction in serum cholesterol—which may be attainable through increased fish consumption as described above—is associated with a 9% reduced risk of nonfatal MI after 2 years of follow-up and 37% reduction after 5 years [53]. However, these findings are not as consistent [8]. It has been hypothesized that omega-3 PUFAs are associated with decreased CHD mortality by increasing the threshold for ventricular fibrillation [54], which may explain why oftentimes associations have been observed for fatal events only. Evidence from animal studies suggests that omega-3 PUFAs reduce the risk of triggered ventricular arrhythmias by a partial stabilization of depolarized ischemic myocytes [8]. This physiological mechanism is consistent with the above hypothesis and clinical evidence showing reductions in CHD mortality among those who eat more omega-3 PUFAs.

8.7 EPIDEMIOLOGICAL EVIDENCE—MEAT INTAKE AND CARDIOVASCULAR DISEASE

The relationship between red and processed meat intake and cardiovascular disease has also been the focus of much nutritional epidemiological research during the past few decades. This is of interest when considering fish intake and cardiovascular disease fish is often consumed instead of, rather than in addition to, red and processed meats. The evidence is conclusive that meat intake has negative cardiovascular consequences. A 2012 systematic review found that processed meats are associated with higher incidence of CHD; each additional serving per day of processed meat (e.g., bacon, hot dogs, salami, sausages, and processed deli meat) was associated with a 42% higher risk of CHD [11].

A 2010 systematic review and meta-analysis of 20 cohort studies, case-control studies, and RCTs investigated the relationship between meat (unprocessed, processed, red, and/or total) and CHD, stroke, and/or diabetes mellitus [11]. The meta-analysis of the four studies that analyzed red meat and CHD found that each 100-g serving per day of red meat was not associated with increased risk of CHD. However, the meta-analysis of five studies analyzing processed meat and CHD found that each 50-g serving per day of processed meat was associated with a 52% increased risk of CHD. Although there are only small amounts of trans fat in red and processed meat, another meta-analysis found that increasing energy intake from trans fat by only 2% was associated with a 23% increased risk of CHD. However, when looking at trans fat from animal studies only, an increased risk has not been detected [1].

Many cohort studies have investigated the relationship between type of dietary fat and CHD. Overall, they have found that while total fat was not associated with CHD, substituting subtypes of dietary fat for carbohydrates show that mono- and polyunsaturated fats, abundant in fish, were associated with significantly decreased risk compared to saturated and trans fat found in red and processed meat [1,55,56] (Fig. 8.2). For example, a prospective study of US men comparing major dietary patterns found that a dietary pattern with high intake of red and processed meat was associated with an increased risk of CHD (other characteristics of this dietary pattern included refined cereals and high-fat dairy products) [57]. Another prospective study found that higher intake of red meat was associated with higher risk of heart failure [58].

Numerous RCTs have investigated by either substituting polyunsaturated fat for saturated fat or substituting carbohydrates for saturated fat. Despite critical limitations such as small sample size and barriers to long-term compliance and contamination, evidence supports that a diet low in saturated fat and low in unsaturated fat is ideal for cardiovascular disease prevention [1]. The Oslo Heart Study [50] provided particularly convincing evidence; 1232 men with high serum cholesterol and normal blood pressure (80% smokers) were assigned to either a smoking cessation program or dietary intervention,

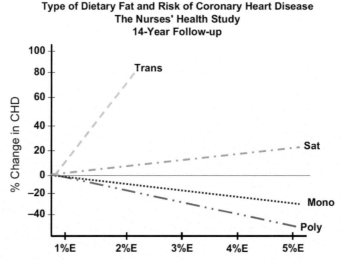

FIGURE 8.2 Multivariate relative risks of coronary heart disease with increasing intakes of specific types of fat compared with the same energy intake from carbohydrate. Data are based on 14 years of follow-up in the Nurses' Health Study. *From Hu FB, Stampfer MJ, Manson JE, Rimm E, Colditz GA, Rosner BA, et al. Dietary fat intakeand the risk of coronary heart disease in women. N Engl J Med 1997;337(21):1491—9; used with permission.*

which included reducing saturated fat and cholesterol intake and a minor increase in polyunsaturated fat intake. After 5 years, serum cholesterol was 13% lower in the intervention group, and incidence of nonfatal MI and fatal CHD was 47% lower [1].

8.8 FISH, MERCURY, AND CARDIOVASCULAR DISEASE

Despite the established health benefits of fish intake, there is concern about the potential adverse effects of mercury from fish; fish is a major source of mercury [9], which, at extremely high levels, has been shown to result in serious health complications including cardiovascular disease [7].

However, a 2012 review of mercury exposure and human health concluded that there is no clear relationship between mercury exposure and cardiovascular disease [59]. A summary of risk-benefit analyses concluded that the health benefits of modest fish consumption significantly outweigh the potential risks [8], including those related to mercury contamination. Mercury content is low in most fish species, moderate in a few (i.e., albacore tuna), and high (according to Federal Drug Administration of the US action level of 1 μg/g) in golden bass, swordfish, shark, and king mackerel from the Gulf of Mexico. US-based studies investigating the relationship between fish intake, mercury content, and cardiovascular disease do not find adverse effects of mercury exposure on CHD, stroke, hypertension, or total cardio-vascular disease [9]. This was true for levels 2.5-fold higher than the US Environmental Protection Agency reference dose [60]. A study adults at high risk of cardiovascular disease from Spain, where fish consumption is very high, also finds no adverse effects [61,62].

Regardless, several studies investigating the relationship between fish intake, mercury exposure, and cardiovascular disease in adults have gener-ated inconsistent results [63–70]. Government agencies and health organiza-tions have identified mercury exposure from fish consumption as an area needing further investigation [7,71–74], and while evidence strongly sug-gests that the mercury content of fish does not negate the cardiovascular ben-efits of fish consumption, a stronger evidence base is needed before official recommendations are made. Furthermore, mercury is unsafe for sensitive populations including pregnant women and young children due to neurologi-cal consequences in developing brains [75], which involve separate risk fac-tors than cardiovascular disease.

8.9   CONCLUSIONS

Evidence strongly suggests that a diet high in fish and low in red and pro-cessed meats is beneficial for cardiovascular disease. Yet there are still many things that we do not know, including the exact physiological effects and molecular mechanisms behind the observed cardiovascular benefits of fish consumption and whether magnitudes of effect and dose-response relationships vary across populations and specific clinical outcomes [8]. Further research is also needed on the potential benefit of ALA present in vegetables and nonfish dietary sources, as this could be a lower cost and more available alternative to EPA and DHA for many parts of the world [8]. Regardless of these areas of uncertainty, it remains clear that making such dietary modifications is para-mount for the prevention of cardiovascular disease.

8.10   RECOMMENDATIONS

It is recommended to include fish every week about three or more servings, including at least one fatty fish.

A serving of fish is equivalent to 125–150 g in raw, which could be an individual fillet or two slices. One small canned fish can also be considered one serving size.

Seafood is also included in these recommendations of fish. Despite hav-ing a bad reputation for its high cholesterol and purine (associated with uric acid and gout) content, seafood is generally a less frequently consumed food, which when eaten is consumed in small amounts. For this reason, the effect it may have on cholesterol in blood is insignificant. In addition, seafood has a low percentage of fat and a similar amount of omega-3 in comparison with that of fish. People who will have to moderate their consumption of seafood are those with high levels of uric acid or who have gout.

If these servings of fish and seafood are consumed weekly, we help reach the recommended ingest of omega-3 fatty acids.

How to reach the recommendation:

- Buy a variety of different fish: cod fish, salmon, tuna, tilapia, snapper, sardines, trout, catfish, etc.
- Try to vary the culinary technology applied to the fish when preparing the dish. There are multiple forms for preparing fish: grilled, baked, fried, breaded, salted, steamed, smoked, microwaved, or as an ingredient in many recipes.
- Canned fish is a healthy and economical source of fish. It is preferable to buy those canned naturally or in olive oil. Canned fish can make for great sandwiches or be added into salads.
- Another option is to buy frozen fish, equally as healthy that you can always have at hand if you cannot buy fresh fish frequently.
- Fish goes well with pasta, rice, and potato dishes, serving as a good source of high-quality protein and healthy fat.
- Fish cooked with beans form a highly nutritious dish, substituting the more frequent combination with meat.
- Fish is a much healthier option instead of cold cuts and other meats typically used in sandwiches.
- Fish hamburgers and meatballs are an acceptable alternative for incorporating fish into a diet, particularly helpful for feeding children fish.
- Include in omelets and scrambled eggs, creating a tastier and nutritious meal.
- Surprise your guests at celebrations and gatherings with fish-based dishes: Fish skewers, grilled fish, seafood salad, caviar appetizers, etc.
- Fish soup serves as a nice, comforting, and healthy dish for cold days.
- Some pates are actually made with fish, which can be used as a spread for bread as a healthy snack.

REFERENCES

[1] Willett WC. Dietary fats and coronary heart disease. J Intern Med 2012;272(1):13−24.
[2] Dyerberg J, Bang HO, Hjorne N. Fatty-acid composition of plasma-lipids in Greenland Eskimos. Am J Clin Nutr 1975;28(9):958−66.
[3] Kagawa Y, Nishizawa M, Suzuki M, Miyatake T, Hamamoto T, Goto K, et al. Eicosapolyenoic acids of serum lipids of Japanese islanders with low incidence of cardiovascular diseases. J Nutr Sci Vitaminol (Tokyo) 1982;28(4):441−53.
[4] Kromhout D, Feskens EJ, Bowles CH. The protective effect of a small amount of fish on coronary heart disease mortality in an elderly population. Int J Epidemiol 1995;24(2):340−5.

[5]   Kato H, Tillotson J, Nichaman MZ, Rhoads GG, Hamilton HB. Epidemiologic studies of coronary heart disease and stroke in Japanese men living in Japan, Hawaii and California. Am J Epidemiol 1973;97(6):372–85.

[6]   Gibbons A. The evolution of diet. Natl Geogr 2014;226:30.

[7]   Mozaffarian D, Rimm EB. Fish intake, contaminants, and human health: evaluating the risks and the benefits. JAMA 2006;296(15):1885–99.

[8]   Mozaffarian D, Wu JH. Omega-3 fatty acids and cardiovascular disease: effects on risk factors, molecular pathways, and clinical events. J Am Coll Cardiol 2011;58 (20):2047–67.

[9]   Mozaffarian D, Shi P, Morris JS, Spiegelman D, Grandjean P, Siscovick DS, et al. Mercury exposure and risk of cardiovascular disease in two U.S. cohorts. N Engl J Med 2011;364(12):1116–25.

[10] Lichtenstein AH. Dietary trans fatty acids and cardiovascular disease risk: past and present. Curr Atheroscler Rep 2014;16(8):433.

[11] Micha R, Wallace SK, Mozaffarian D. Red and processed meat consumption and risk of incident coronary heart disease, stroke, and diabetes mellitus: a systematic review and meta-analysis. Circulation 2010;121(21):2271–83.

[12] Harris WS, Bulchandani D. Why do omega-3 fatty acids lower serum triglycerides? Curr Opin Lipidol 2006;17(4):387–93.

[13] Moslehi N, Ehsani B, Mirmiran P, Hojjat P, Azizi F. Association of dietary proportions of macronutrients with visceral adiposity index: non-substitution and iso-energetic substitution models in a prospective study. Nutrients 2015;7(10):8859–70.

[14] Oshima Y, Rin S, Kita H, Hiramoto Y, Morimoto S, Okunaga A, et al. The frequency of fish-eating could negatively associate with visceral adiposity in those who eat moderately. J Nutr Sci Vitaminol (Tokyo) 2015;61(5):426–31.

[15] Mensink RP, Zock PL, Katan MB, Hornstra G. Effect of dietary cis and trans fatty acids on serum lipoprotein[a] levels in humans. J Lipid Res 1992;33(10):1493–501.

[16] Gebauer SK, Chardigny JM, Jakobsen MU, Lamarche B, Lock AL, Proctor SD, et al. Effects of ruminant trans fatty acids on cardiovascular disease and cancer: a comprehensive review of epidemiological, clinical, and mechanistic studies. Adv Nutr 2011;2 (4):332–54.

[17] Tranchida F, Tchiakpe L, Rakotoniaina Z, Deyris V, Ravion O, Hiol A. Long-term high fructose and saturated fat diet affects plasma fatty acid profile in rats. J Zhejiang Univ Sci B 2012;13(4):307–17.

[18] Djousse L, Gaziano JM, Buring JE, Lee IM. Egg consumption and risk of type 2 diabetes in men and women. Diabetes Care 2009;32(2):295–300.

[19] Portha B, Giroix MH, Cros JC, Picon L. Diabetogenic effect of N-nitrosomethylurea and N-nitrosomethylurethane in the adult rat. Ann Nutr Aliment 1980;34(5-6):1143–51.

[20] McGrowder D, Ragoobirsingh D, Dasgupta T. Effects of S-nitroso-N-acetyl-penicillamine administration on glucose tolerance and plasma levels of insulin and glucagon in the dog. Nitric Oxide 2001;5(4):402–12.

[21] Lloyd-Jones DM, Larson MG, Leip EP, Beiser A, D'Agostino RB, Kannel WB, et al. Framingham Heart Study. Lifetime risk for developing congestive heart failure: the Framingham Heart Study. Circulation 2002;106(24):3068–72.

[22] Nichols GA, Hillier TA, Erbey JR, Brown JB. Congestive heart failure in type 2 diabetes: prevalence, incidence, and risk factors. Diabetes Care 2001;24(9):1614–19.

[23]   Leung AA, Eurich DT, Lamb DA, Majumdar SR, Johnson JA, Blackburn DF, et al. Risk of heart failure in patients with recent-onset type 2 diabetes: population-based cohort study. J Card Fail 2009;15(2):152−7.

[24]   Rouhani MH, Salehi-Abargouei A, Surkan PJ, Azadbakht L. Is there a relationship between red or processed meat intake and obesity? A systematic review and meta-analysis of observational studies. Obes Rev 2014;15(9):740−8.

[25]   Hariri N, Gougeon R, Thibault L. A highly saturated fat-rich diet is more obesogenic than diets with lower saturated fat content. Nutr Res 2010;30(9):632−43.

[26]   Liao FH, Liou TH, Shieh MJ, Chien YW. Effects of different ratios of monounsaturated and polyunsaturated fatty acids to saturated fatty acids on regulating body fat deposition in hamsters. Nutrition 2010;26(7-8):811−17.

[27]   Hansen CP, Berentzen TL, Halkjaer J, Tjonneland A, Sorensen TI, Overvad K, et al. Intake of ruminant trans fatty acids and changes in body weight and waist circumference. Eur J Clin Nutr 2012;66(10):1104−9.

[28]   Mozaffarian D, Aro A, Willett WC. Health effects of trans-fatty acids: experimental and observational evidence. Eur J Clin Nutr 2009;63(Suppl 2)):S5−21.

[29]   Sanders PW. Vascular consequences of dietary salt intake. Am J Physiol Renal Physiol 2009;297(2):F237−43.

[30]   Forstermann U. Oxidative stress in vascular disease: causes, defense mechanisms and potential therapies. Nat Clin Pract Cardiovasc Med 2008;5(6):338−49.

[31]   Sacks FM, Svetkey LP, Vollmer WM, Appel LJ, Bray GA, Harsha D, et al. Effects on blood pressure of reduced dietary sodium and the Dietary Approaches to Stop Hypertension (DASH) diet. DASH-Sodium Collaborative Research Group. N Engl J Med 2001;344(1):3−10.

[32]   Townsend RR, Kapoor S, McFadden CB. Salt intake and insulin sensitivity in healthy human volunteers. Clin Sci (Lond) 2007;113(3):141−8.

[33]   He FJ, MacGregor GA. Effect of modest salt reduction on blood pressure: a meta-analysis of randomized trials. Implications for public health. J Hum Hypertens 2002;16(11):761−70.

[34]   Wang L, Manson JE, Buring JE, Sesso HD. Meat intake and the risk of hypertension in middle-aged and older women. J Hypertens 2008;26(2):215−22.

[35]   Miura K, Greenland P, Stamler J, Liu K, Daviglus ML, Nakagawa H. Relation of vegetable, fruit, and meat intake to 7-year blood pressure change in middle-aged men: the Chicago Western Electric Study. Am J Epidemiol 2004;159(6):572−80.

[36]   Steffen LM, Kroenke CH, Yu X, Pereira MA, Slattery ML, Van Horn L, et al. Associations of plant food, dairy product, and meat intakes with 15-y incidence of elevated blood pressure in young black and white adults: the Coronary Artery Risk Development in Young Adults (CARDIA) Study. Am J Clin Nutr 2005;82(6):1169−77 quiz 363−4.

[37]   Tzoulaki I, Brown IJ, Chan Q, Van Horn L, Ueshima H, Zhao L, et al. Relation of iron and red meat intake to blood pressure: cross sectional epidemiological study. BMJ 2008;337:a258.

[38]   Stamler J, Stamler R, Neaton JD. Blood pressure, systolic and diastolic, and cardiovascular risks. US population data. Arch Intern Med 1993;153(5):598−615.

[39]   Norell SE, Ahlbom A, Feychting M, Pedersen NL. Fish consumption and mortality from coronary heart disease. Br Med J (Clin Res Ed) 1986;293(6544):426.

[40]   Dolecek TA. Epidemiological evidence of relationships between dietary polyunsaturated fatty acids and mortality in the multiple risk factor intervention trial. Proc Soc Exp Biol Med 1992;200(2):177−82.

[41] Jakobsen MU, O'Reilly EJ, Heitmann BL, Pereira MA, Balter K, Fraser GE, et al. Major types of dietary fat and risk of coronary heart disease: a pooled analysis of 11 cohort studies. Am J Clin Nutr 2009;89(5):1425−32.

[42] Key TJ, Fraser GE, Thorogood M, Appleby PN, Beral V, Reeves G, et al. Mortality in vegetarians and nonvegetarians: detailed findings from a collaborative analysis of 5 prospective studies. Am J Clin Nutr 1999;70(3 Suppl):516S−24S.

[43] Women's Health Initiative Study Group. Dietary adherence in the Women's Health Initiative Dietary Modification Trial. J Am Diet Assoc 2004;104(4):654−8.

[44] Multiple risk factor intervention trial. Risk factor changes and mortality results. Multiple Risk Factor Intervention Trial Research Group. JAMA 1982;248(12):1465−77.

[45] Burr ML, Fehily AM, Gilbert JF, Rogers S, Holliday RM, Sweetnam PM, et al. Effects of changes in fat, fish, and fibre intakes on death and myocardial reinfarction: diet and reinfarction trial (DART). Lancet 1989;2(8666):757−61.

[46] Burr ML, Ashfield-Watt PA, Dunstan FD, Fehily AM, Breay P, Ashton T, et al. Lack of benefit of dietary advice to men with angina: results of a controlled trial. Eur J Clin Nutr 2003;57(2):193−200.

[47] Kromhout D, Giltay EJ, Geleijnse JM, Alpha Omega Trial Group. n-3 fatty acids and cardiovascular events after myocardial infarction. N Engl J Med 2010;363(21):2015−26.

[48] Galan P, Kesse-Guyot E, Czernichow S, Briancon S, Blacher J, Hercberg S, SU.FOL. OM3 Collaborative Group. Effects of B vitamins and omega 3 fatty acids on cardiovascular diseases: a randomised placebo controlled trial. BMJ 2010;341:c6273.

[49] Rauch B, Schiele R, Schneider S, Diller F, Victor N, Gohlke H, et al. OMEGA, a randomized, placebo-controlled trial to test the effect of highly purified omega-3 fatty acids on top of modern guideline-adjusted therapy after myocardial infarction. Circulation 2010;122(21):2152−9.

[50] Hjermann I, Velve Byre K, Holme I, Leren P. Effect of diet and smoking intervention on the incidence of coronary heart disease. Report from the Oslo Study Group of a randomised trial in healthy men. Lancet 1981;2(8259):1303−10.

[51] de Lorgeril M, Renaud S, Mamelle N, Salen P, Martin JL, Monjaud I, et al. Mediterranean alpha-linolenic acid-rich diet in secondary prevention of coronary heart disease. Lancet 1994;343(8911):1454−9.

[52] Guasch-Ferre M, Babio N, Martinez-Gonzalez MA, Corella D, Ros E, Martin-Pelaez S, et al. Dietary fat intake and risk of cardiovascular disease and all-cause mortality in a population at high risk of cardiovascular disease. Am J Clin Nutr 2015;102(6):1563−73.

[53] Law MR, Wald NJ, Thompson SG. By how much and how quickly does reduction in serum cholesterol concentration lower risk of ischaemic heart disease? BMJ 1994;308 (6925):367−72.

[54] Leaf A. Omega-3 fatty acids and prevention of ventricular fibrillation. Prostaglandins Leukot Essent Fatty Acids 1995;52(2-3):197−8.

[55] Hu FB, Stampfer MJ, Manson JE, Rimm E, Colditz GA, Rosner BA, et al. Dietary fat intake and the risk of coronary heart disease in women. N Engl J Med 1997;337(21):1491−9.

[56] Hjermann I, Holme I, Leren P. Oslo Study Diet and Antismoking Trial. Results after 102 months. Am J Med 1986;80(2A):7−11.

[57] Hu FB, Rimm EB, Stampfer MJ, Ascherio A, Spiegelman D, Willett WC. Prospective study of major dietary patterns and risk of coronary heart disease in men. Am J Clin Nutr 2000;72(4):912−21.

[58] Ashaye A, Gaziano J, Djousse L. Red meat consumption and risk of heart failure in male physicians. Nutr Metab Cardiovasc Dis 2011;21(12):941−6.

[59] Karagas MR, Choi AL, Oken E, Horvat M, Schoeny R, Kamai E, et al. Evidence on the human health effects of low-level methylmercury exposure. Environ Health Perspect 2012;120(6):799−806.

[60] Mozaffarian D, Shi P, Morris JS, Grandjean P, Siscovick DS, Spiegelman D, et al. Mercury exposure and risk of hypertension in US men and women in 2 prospective cohorts. Hypertension 2012;60(3):645−52.

[61] FAOSTAT, Food supply—Livestock and fish primary equivalent, Europe, S.D. Food and Agriculture Organization of the United Nations. 2011.

[62] Downer MK, Martinez-Gonzalez MA, Gea A, Stampfer M, Warnberg J, Ruiz-Canela M, et al. Mercury exposure and risk of cardiovascular disease: a nested case-control study in the PREDIMED (PREvention with MEDiterranean Diet) study. BMC Cardiovasc Disord 2017;17(1):9.

[63] Guallar E, Sanz-Gallardo MI, van't Veer P, Bode P, Aro A, Gomez-Aracena J, et al. Mercury, fish oils, and the risk of myocardial infarction. N Engl J Med 2002;347 (22):1747−54.

[64] Virtanen JK, Voutilainen S, Rissanen TH, Mursu J, Tuomainen TP, Korhonen MJ, et al. Mercury, fish oils, and risk of acute coronary events and cardiovascular disease, coronary heart disease, and all-cause mortality in men in eastern Finland. Arterioscler Thromb Vasc Biol 2005;25(1):228−33.

[65] Yoshizawa K, Rimm EB, Morris JS, Spate VL, Hsieh CC, Spiegelman D, et al. Mercury and the risk of coronary heart disease in men. N Engl J Med 2002;347 (22):1755−60.

[66] Wennberg M, Bergdahl IA, Stegmayr B, Hallmans G, Lundh T, Skerfving S, et al. Fish intake, mercury, long-chain n-3 polyunsaturated fatty acids and risk of stroke in northern Sweden. Br J Nutr 2007;98(5):1038−45.

[67] Ahlqwist M, Bengtsson C, Lapidus L, Gergdahl IA, Schutz A. Serum mercury concentration in relation to survival, symptoms, and diseases: results from the prospective population study of women in Gothenburg, Sweden. Acta Odontol Scand 1999;57 (3):168−74.

[68] Hallgren CG, Hallmans G, Jansson JH, Marklund SL, Huhtasaari F, Schutz A, et al. Markers of high fish intake are associated with decreased risk of a first myocardial infarction. Br J Nutr 2001;86(3):397−404.

[69] Kim YN, Kim YA, Yang AR, Lee BH. Relationship between Blood Mercury Level and Risk of Cardiovascular Diseases: Results from the Fourth Korea National Health and Nutrition Examination Survey (KNHANES IV) 2008-2009. Prev Nutr Food Sci 2014;19 (4):333−42.

[70] Turunen AW, Jula A, Suominen AL, Mannisto S, Marniemi J, Kiviranta H, et al. Fish consumption, omega-3 fatty acids, and environmental contaminants in relation to low-grade inflammation and early atherosclerosis. Environ Res 2013;120:43−54.

[71] Rice DC. The US EPA reference dose for methylmercury: sources of uncertainty. Environ Res 2004;95(3):406−13.

[72] Konig A, Bouzan C, Cohen JT, Connor WE, Kris-Etherton PM, Gray GM, et al. A quantitative analysis of fish consumption and coronary heart disease mortality. Am J Prev Med 2005;29(4):335−46.

[73] Keating M.M., Rice G., Bullock O., Ambrose R., Swartout J., Nichols J. Mercury Study Report to Congress. 1997, United States Environmental Protection Agency: Washington, D.C. p. 95. 1997.

[74] Report of the Joint FAO/WHO Expert Consultation on the Risks and Benefits of Fish Consumption. Food and Agriculture Organization of the United Nations; Geneva. Rome, Italy: World Health Organization; 2011.

[75] McGuire S. U.S. Department of Agriculture and U.S. Department of Health and Human Services, Dietary Guidelines for Americans, 2010. 7th Edition, Washington, DC: U.S. Government Printing Office, January 2011. Adv Nutr 2011;2(3):293−4.

Chapter 9

Red Wine Moderate Consumption and at Mealtimes

Alfredo Gea and Ana Sánchez-Tainta

9.1 INTRODUCTION

Alcohol is sometimes considered as a drug since some people use it seeking the psychoactive effects of heavy alcohol intake. Moreover, as any other drink, alcohol intake can also be thought as part of the diet. In general, alcoholic beverages contain mainly water, alcohol, and carbohydrates, and also some micronutrients, that contribute to the overall dietary pattern. However, alcohol intake may have detrimental effects on health that frequently overcome its potential benefits, and those effects depend on the characteristics of a person. Therefore, when we discuss alcohol intake, two principles should rule our recommendations: the precautionary principle, and the stratification of the message.

There are different foods or food groups that are important components of the Mediterranean Diet, and they are presented in this work. One of those components is red wine. Out of all the alcoholic beverages, red wine is the one consumed within the Mediterranean Diet. Land and weather in Mediterranean countries are appropriate to grow grapevines, and therefore vineyards have been present in those countries for millennia. However, for the sake of truth, we must say that not all the Mediterranean countries shared this characteristic: in southern-eastern countries of the Mediterranean basin, traditionally Muslim, most people usually abstain from alcohol, in accordance with their precepts.

Red wine is the product of the fermentation of red-colored grapes. Sugars in grapes turn into alcohol. The red color is due to some substances in the skin of grapes. And if there is a difference between the effect of red wine and the effect of white wine, it would be determined by some molecules in the skin of grapes that are in red wine but not in white wine, or by-products of the fermentation of those substances.

However, as previously stated in this work, Mediterranean Diet is not only the sum of a series of foods but also the circumstances surrounding the

The Prevention of Cardiovascular Disease through The Mediterranean Diet.
DOI: http://dx.doi.org/10.1016/B978-0-12-811259-5.00009-3
151

act of eating and drinking. In the Mediterranean culture meals used to take place at the table, with an important social component, and at a friendly and calm environment, that also affect the way of eating. All these circumstances and some others also apply for the so- called *Mediterranean alcohol-drinking pattern*, or the *Mediterranean way of drinking*. Traditionally, wine was consumed every day in small amounts, without days of excess consumption, and always with meals. In this scenario, people considered alcohol consumption as part of the diet. Nowadays, however, this drinking pattern is less prevalent since people, especially at young ages, tend to drink higher quantities in shorter time periods, and usually drink distilled beverages. This drinking pattern is known as *binge-drinking*, and it may lead to drunkenness. In this other scenario, people use alcohol as a drug, seeking the psychoactive effects of alcohol, regardless of whether people get drunk or not.

The binge-drinking pattern and the Mediterranean alcohol-drinking pattern are opposite to some extent. Regarding the quantity, it is high in the binge-drinking pattern and moderate in the Mediterranean pattern. In terms of frequency, alcohol intake is daily in the Mediterranean pattern, and binge-drinking is usually occasional, but it can occur more frequently. And finally, binge-drinking occasions do not usually take place at mealtimes. Regarding the type of beverage, people can binge-drink with any alcoholic beverage, but spirits-drinkers are more likely to binge-drink than wine-drinkers.

These two drinking patterns have very different effects on health. And of course, those effects are due not only to the quantity, but also to the rest of characteristics of alcohol consumption. In general, moderate wine consumption at mealtimes is associated with a lower risk of cardiovascular disease, diabetes, hypertension, depression, and total mortality, compared to the abstention.

9.2 SCIENTIFIC EVIDENCE

When we look at the scientific literature, we should take into account a very important methodological issue. Most of the evidence we have is focused on one single aspect of alcohol drinking, usually the quantity consumed. However, as we have already mentioned, alcohol drinking is a multidimensional exposure, and each of the dimensions may have its own effect, and may modify the effects of the other dimensions. For instance, the effect would not be the same for seven glasses of wine in a single occasion or one per day, even though the weekly quantity is the same. Therefore, our conclusions should rely more in evidence about the overall drinking pattern, instead of focusing on a single dimension.

9.2.1 Mortality

There is plenty of literature that relates moderate alcohol consumption with a lower risk of cancer mortality [1], cardiovascular mortality [2], and total mortality [3], and that presents a higher risk of mortality among heavy drinkers. Moreover, assessing the overall drinking pattern, the Mediterranean alcohol-drinking pattern was associated with a lower risk of mortality, compared to drinking out of that pattern, and also with the abstention [4]. This is probably due to the benefits on cardiovascular health and the nonelevated risk on cancer and death by external causes [5]. In terms of wine, moderate consumption would be 2−3 glasses/day for men and 1−2 glass/day for women (a glass being 100−150 mL approximately)—there are some differences between men and women, and, in general terms, women should drink less than men, because of the different body composition, different metabolism of alcohol, and some other differences that make blood alcohol levels differ, given the same quantity of alcohol intake—it is important to notice that the effect is modified by each of the different aspects in the *Mediterranean alcohol-drinking pattern*. For instance, drinkers outside of meals had a higher risk of mortality in a study by Trevisan et al. [6] compared with those who consumed alcohol with meals. Moreover, among usually moderate drinkers, episodic heavy drinking has been related to a more than twofold higher risk of mortality [7]. And regarding the type of beverage consumed, although there is still some controversy in the scientific community about the causes, moderate wine-drinkers are at lower risk of mortality than moderate other-beverages-drinkers [8].

9.2.2 Cardiovascular Diseases

Moderate alcohol consumption has been consistently associated with a lower risk of coronary heart disease than abstainers [9]. The main protective effect has been attributed to alcohol itself. The preferred beverage seems not to affect the effect of alcohol, although not all authors agree. However, other characteristics of the drinking pattern may be important, as the frequency [10] or the avoidance of binge-drinking. Nevertheless, more studies are needed to confirm the role of the overall drinking pattern [11]. On the other hand, moderate alcohol consumption has been related to a lower risk of ischemic stroke, while there is a higher risk among heavy drinkers. And although less prevalent, hemorrhagic stroke occurs in a higher rate among alcohol drinkers, with a linear dose−response relationship [12]. Some possible mechanisms are the effects on lipids, platelet coagulability, and inflammation.

We also know that heavy alcohol intake is associated with a higher risk of hypertension. However, moderate alcohol intake seems to reduce the risk of hypertension, mainly among women [13]. The relationship between the overall drinking pattern and hypertension has not been assessed, but some studies proposed that binge-drinking and beer and spirits consumption might be related to a higher risk of hypertension than not bingeing and wine consumption [14,15].

9.2.3 Diabetes

Several studies have assessed the relationship between alcohol intake and the incidence of diabetes, and recent meta-analyses concluded that moderate alcohol intake was associated with a lower risk of developing type 2 diabetes. Moreover, that association was stronger among people that consumed moderate quantities of alcohol almost every day [16,17], and also among wine drinkers [18]. This effect may be mediated through higher insulin sensitivity as a consequence of alcohol intake.

9.2.4 Depression

Some researchers linked depression to cardiovascular disease, hypothesizing that both may share common pathological pathways. In the light of this hypothesis, alcohol intake and specifically moderate wine consumption, has been related to a lower risk of developing depression [19]. However, heavy alcohol intake or binge-drinking have been related to a higher risk of depression, and alcohol should not be used as a treatment of depression.

9.2.5 Weight

As mentioned before, alcoholic beverages contribute to overall diet mainly with alcohol and carbohydrates (and some micronutrients). Energy content in 1 g of alcohol is 7.1 kcal. But those are called "empty calories" because it is oxidized before fat and carbohydrates since it cannot be stored. Therefore, heavy alcohol consumption has been associated with a higher risk of obesity or a higher weight gain [20]. However, moderate wine consumption may not greatly affect the energy balance and, on the other hand may exert a benefit through the antiinflammatory substances.

9.2.6 Cancer

Ethanol and its metabolite acetaldehyde are considered carcinogenic. In high doses, alcohol intake is associated with a higher risk of cancer of oral cavity, pharynx, esophagus, colon-rectum, liver, larynx, female breast cancer, stomach, liver, gallbladder, pancreas, and lung cancer [21]. Although this relationship is not that well established for moderate alcohol intake, these associations are one of the main reasons to be cautious with the recommendation of alcohol intake. However, there are no studies assessing the effect on the incidence of cancer of the overall drinking pattern. It seems reasonable that if alcohol is consumed with meals, the potential damage to the gastrointestinal tract would be way lower. Moreover, antioxidants present in red wine may counteract the effect of alcohol. And additionally, blood alcohol concentration is smaller if alcohol is consumed in small quantities and every day instead of bingeing. Nevertheless, more research is needed.

9.2.7 Other Health Effects

There are some other effects of moderate alcohol consumption on different health outcomes. Heavy drinking leads to an increased risk of suicide, involvement in violent situations, traffic injuries and other accidental injuries, or liver cirrhosis. However, moderate alcohol intake is not related to most of the previous. Moreover, the Mediterranean alcohol-drinking pattern may exert an effect on the social and psychological dimensions, not because of alcohol but due to the surrounding circumstances of that drinking pattern.

9.3 RECOMMENDATIONS

As we mentioned previously, when we discuss alcohol intake, two principles should rule our recommendations: the precautionary principle, and the stratification of the message.

The effect of alcohol on health is the sum of all the possible consequences, beneficial or harmful, of the alcohol-drinking pattern. All dimensions of the overall drinking pattern are important for the total effect. And though there is more research needed, the Mediterranean alcohol-drinking pattern seems to be a good choice [22]. However, there are some potential harmful effects that should be avoided. For that reason, group of people with a higher probability of those risks are not recommended to drink, not even in a moderate way, based on the precautionary principle.

Alcohol intake is not recommended for pregnant or breastfeeding women. There is evidence that heavy alcohol intake affects the baby, but as we do not know if there is a safe dose, we should recommend them not to drink. In the same way, adolescents and young adults would not have the benefits of alcohol consumption but may suffer the harmful effects. Their baseline hazard of the mentioned chronic diseases is very low; therefore, the potential benefits are almost zero. So alcohol consumption should not be recommended to adolescents and young adults. There are groups of people that empirically know they do not tolerate alcohol as good as the rest of the population. These people should not drink much since they may be slow metabolizers or intolerant to alcohol and have a higher risk associated to alcohol intake. Another group of people that should not drink at all is drivers. Any alcohol intake affects the skills needed to drive or to work with different machines.

And finally, we should not forget that alcohol intake is one of the main risk factors that contribute to the burden of disease in the world. Therefore, a very important recommendation would be to avoid heavy drinking or binge-drinking, maybe changing that drinking pattern to a more Mediterranean way of drinking. However, abstainers should not start drinking since some of them may end up drinking in harmful levels, according to the *slippery slope* theory. Future research will answer some of the remaining questions, and more precise recommendations will be given in the future.

REFERENCES

[1] Jin M, Cai S, Guo J, Zhu Y, Li M, Yu Y, et al. Alcohol drinking and all cancer mortality: a meta-analysis. Ann Oncol 2013;24(3):807−16.

[2] Costanzo S, Di Castelnuovo A, Donati MB, Iacoviello L, de Gaetano G. Alcohol consumption and mortality in patients with cardiovascular disease: a meta-analysis. J Am Coll Cardiol 2010;55(13):1339−47.

[3] Di Castelnuovo A, Costanzo S, Bagnardi V, Donati MB, Iacoviello L, de Gaetano G. Alcohol dosing and total mortality in men and women: an updated meta-analysis of 34 prospective studies. Arch Intern Med 2006;166(22):2437−45.

[4] Gea A, Bes-Rastrollo M, Toledo E, Garcia-Lopez M, Beunza JJ, Estruch R, et al. Mediterranean alcohol-drinking pattern and mortality in the SUN (Seguimiento Universidad de Navarra) Project: a prospective cohort study. Br J Nutr 2014;111(10):1871−80.

[5] Giacosa A, Barale R, Bavaresco L, Faliva MA, Gerbi V, La Vecchia C, et al. Mediterranean way of drinking and longevity. Crit Rev Food Sci Nutr 2016;56 (4):635−40.

[6] Trevisan M, Schisterman E, Mennotti A, Farchi G, Conti S, Risk Factor and Life Expectancy Research Group. Drinking pattern and mortality: the Italian Risk Factor and Life Expectancy pooling project. Ann Epidemiol 2001;11(5):312−19.

[7] Holahan CJ, Schutte KK, Brennan PL, Holahan CK, Moos RH. Episodic heavy drinking and 20-year total mortality among late-life moderate drinkers. Alcohol Clin Exp Res 2014;38(5):1432−8.

[8] Klatsky AL, Friedman GD, Armstrong MA, Kipp H. Wine, liquor, beer, and mortality. Am J Epidemiol 2003;158(6):585−95.

[9] Corrao G, Rubbiati L, Bagnardi V, Zambon A, Poikolainen K. Alcohol and coronary heart disease: a meta-analysis. Addiction 2000;95(10):1505−23.

[10] Mukamal KJ, Conigrave KM, Mittleman MA, Camargo Jr CA, Stampfer MJ, Willett WC, et al. Roles of drinking pattern and type of alcohol consumed in coronary heart disease in men. N Engl J Med 2003;348(2):109−18.

[11] Hernandez-Hernandez A, Gea A, Ruiz-Canela M, Toledo E, Beunza JJ, Bes-Rastrollo M, et al. Mediterranean alcohol-drinking pattern and the incidence of cardiovascular disease and cardiovascular mortality: The SUN Project. Nutrients 2015;7(11):9116−26.

[12] Patra J, Taylor B, Irving H, Roerecke M, Baliunas D, Mohapatra S, et al. Alcohol consumption and the risk of morbidity and mortality for different stroke types—a systematic review and meta-analysis. BMC Public Health 2010;10:258.

[13] Briasoulis A, Agarwal V, Messerli FH. Alcohol consumption and the risk of hypertension in men and women: a systematic review and meta-analysis. J Clin Hypertens (Greenwich) 2012;14(11):792−8.

[14] Abramson JL, Lewis C, Murrah NV. Relationship of self-reported alcohol consumption to ambulatory blood pressure in a sample of healthy adults. Am J Hypertens 2010;23 (9):994−9.

[15] Núñez-Córdoba JM, Martínez-González MA, Bes-Rastrollo M, Toledo E, Beunza JJ, Alonso A. Alcohol consumption and the incidence of hypertension in a Mediterranean cohort: the SUN study. Rev Esp Cardiol 2009;62(6):633−41.

[16] Sato KK, Hayashi T, Harita N, Koh H, Maeda I, Endo G, et al. Relationship between drinking patterns and the risk of type 2 diabetes: the Kansai Healthcare Study. J Epidemiol Community Health 2012;66(6):507−11.

[17] Heianza Y, Arase Y, Saito K, Tsuji H, Fujihara K, Hsieh SD, et al. Role of alcohol drinking pattern in type 2 diabetes in Japanese men: the Toranomon Hospital Health Management Center Study 11 (TOPICS 11). Am J Clin Nutr 2013;97(3):561−8.

[18] Huang J, Wang X, Zhang Y. Specific types of alcoholic beverage consumption and risk of type 2 diabetes: a systematic review and meta-analysis. J Diabetes Investig 2017;8 (1):56−68.

[19] Gea A, Beunza JJ, Estruch R, Sánchez-Villegas A, Salas-Salvadó J, Buil-Cosiales P, et al. Alcohol intake, wine consumption and the development of depression: the PREDIMED study. BMC Med 2013;11:192.

[20] Sayon-Orea C, Martinez-Gonzalez MA, Bes-Rastrollo M. Alcohol consumption and body weight: a systematic review. Nutr Rev 2011;69(8):419−31.

[21] Bagnardi V, Rota M, Botteri E, Tramacere I, Islami F, Fedirko V, et al. Alcohol consumption and site-specific cancer risk: a comprehensive dose-response meta-analysis. Br J Cancer 2015;112(3):580−93.

[22] Boban M, Stockley C, Teissedre PL, Restani P, Fradera U, Stein-Hammer C, et al. Drinking pattern of wine and effects on human health: why should we drink moderately and with meals? Food Funct 2016;7(7):2937−42.

Chapter 10

The Mediterranean Lifestyle: Not Only Diet But Also Socializing

Ignacio Ara

10.1 INTRODUCTION

According to the Global Health Observatory of the World Health Organization (WHO) [1], several Mediterranean countries including Spain, Italy, France, Malta, and Greece are among those with the highest life expectancy figures worldwide. Besides, large observational prospective epidemiological studies support the benefits of the Mediterranean dietary pattern for increasing life expectancy, reducing the risk of major chronic disease, and improving quality of life and well-being [2].

Cardiovascular diseases (CVDs) are the leading cause of death in the EU. They account for 51.88 deaths per 100,000 individuals in the population aged under 65 years in the EU27.

Northern European countries—particularly Germany and the United Kingdom—have reported exceptionally high rates of CVD. Southern European countries such as Italy and France have reported relatively low age-standardized death rates from ischemic heart disease in the last 25 years compared to the rest of Europe [3]. The north—south gradient in myocardial infarction and coronary death rates in western European regions was documented by the WHO MONICA Project in the 1990s and has been attributed in part to the Mediterranean Diet.

Moreover, some Mediterranean countries not only show the highest life expectancy rates and lowest mortality rates from-all-causes in Europe (especially in relation to ischemic heart disease and stroke), but also are among those with lowest cancer rates (especially in breast and cervical cancer in women and prostate cancer in men) [4].

These figures can probably not be explained by just the effects of the composition of the Mediterranean Diet. Besides the effectiveness and quality of their different health systems that ensure universal access to quality care

The Prevention of Cardiovascular Disease through The Mediterranean Diet.
DOI: http://dx.doi.org/10.1016/B978-0-12-811259-5.00010-X

for their citizens, there probably exists a *"Mediterranean lifestyle,"* which exerts a strong influence on the people who live in this area and participate in these routines. Thus, beyond the Mediterranean Diet, a *"Mediterranean lifestyle,"* fostered by the climatic and environmental conditions and that includes cultural and social factors such as a high degree of conviviality must be taken into account and considered in order to better understand this complex process. Mental well-being, socialization, social support, and stress management are important components of this phenomenon.

10.2 EXPLORING CARDIOVASCULAR RISK FACTORS IN THE MEDITERRANEAN COUNTRIES: POTENTIAL CONSEQUENCES

Figures related to the most common cardiovascular risk factors (obesity, type 2 diabetes, physical inactivity, tobacco, alcohol consumption, etc.) clearly indicate that Mediterranean countries are unexpectedly low in the ranking when compared to other European countries. As an example, obesity rates are higher in southern European countries compared to Northern European countries (except for the United Kingdom). Surprisingly, this situation appears not to be linked either to higher mortality rates from-all-causes, or to a higher CVD incidence (ischemic heart disease, stroke, etc.) in the southern European countries. On the contrary, as was mentioned before, Mediterranean countries are among those with the lowest disease rates, including the lowest cancer rates (including breast, cervical, and prostate cancers) as well as fostering longer lives. Thus, despite the fact that cardiovascular risk factors seem to be considerably worse than in the rest of the European countries, a higher life expectancy and reduced mortality rates are present in these countries, according to the latest report of the European Commission [4]. On the contrary, one of the few health aspects that seem to be worse in the southern European countries when compared to the rest of Europe is the incidence of dementia. Italy, Greece, France, and Spain were in the top five European countries in the estimated prevalence of dementia per 1000 people in 2015. As is well known, dementia is a common neurological disorder (increasingly prevalent especially among the elderly population), which is linked to heart disease in all its forms, including coronary artery disease, myocardial infarction, atrial fibrillation, valvular disease, and heart failure [5]. Heart disease has been considered as a risk factor for dementia, and its resultant vascular insufficiency has the potential to impair function in other organs, including the brain that might not manifest until the later years of life. Thus, it might be the case that this increased incidence of dementia present in those Mediterranean countries might be related to their exceptional survival rates (the prevalence of dementia across EU countries ranges between 20% and 40% in the over eighties) and the presence of negative cardiovascular risk factors in those countries.

10.3 CLIMATE CONDITIONS AND OPPORTUNITIES FOR A HEALTHY LIFESTYLE

Most parts of southern Europe have a Mediterranean climate characterized by the mildest winters on the European continent, especially in southern and eastern countries of Europe (Spain, Malta, etc.) experiencing distinct wet and dry seasons, with prevailing hot and dry conditions during the summer months. According to the Köppen climate classification [6], this climate is characterized by "cool dry-summer climates" that include: (1) an average temperature above $-3°C$ (27°F), but below 18°C (64°F), (2) a precipitation of less than 30 mm (1.2 in) with less than one-third that of the wettest winter month, and (3) an average temperature in the warmest month above 22°C (72°F). The lands around the Mediterranean Sea form one of the largest areas where this climate type is present, although it can also be found in most of coastal California, in parts of western and southern Australia, in southwestern South Africa, sections of central Asia, and in central Chile.

These special weather conditions make these places of special interest as tourist destinations. In fact, the EU is a major tourist destination, as proven by the fact that five Member States are among the world's top ten destinations for holidaymakers including four with a Mediterranean climate (France, Spain, Italy, and Turkey) according to the 2016 report of the World Tourism Organization (UNWTO) the United Nations specialized agency [7].

10.4 WHY CAN ALL THESE CLIMATE CONDITIONS BE OF INTEREST WHEN STUDYING CARDIOVASCULAR RISK?

Apart from the nutritional aspects associated with these specific climate conditions, the latitude and the longitude of the Mediterranean Sea ($\sim$4.553127, 18.048012) is of special relevance in relation to the availability of daylight. In addition to the psychological effect of strong solar light on human wellbeing, there has been growing evidence for the role of vitamin D in extraskeletal health, including beneficial effects on the cardiovascular system. Epidemiological data suggest that optimal vitamin D status is important for CVD prevention, as proved by published randomized controlled trials on the relation of vitamin D with blood pressure and risk of CVD [8]. Daylight exposure and vitamin D intake in the majority of the Mediterranean countries are abundant for maintaining an adequate vitamin D status.

Moreover, this type of climate offers many opportunities for a healthy lifestyle; one of the major advantages of which is the possibility of performing active outdoor activities during almost the whole year.

In relation to that, it is well known that regular physical activity elicits multiple health benefits in the prevention and management of chronic diseases, including CVD. Some of the known benefits of regular physical activity are associated with lower blood pressure and healthier lipid profiles. As an example, data from the National Health and Nutrition Examination Survey (NHANES) [9] assessed the risks for all-cause and CVD mortality associated with aerobic physical and muscle-strengthening activities. Data from this study showed that in comparison with participants who were physically inactive, the adjusted hazard ratio (HR) for all-cause mortality was 0.64 [95% confidence interval (CI): 0.52−0.79] among those who were physically active (engaging in ≥ 150 min/week of the equivalent moderate-intensity physical activity) and 0.72 (95% CI: 0.54−0.97) among those who were insufficiently active (engaging in >0 to <150 min/week of the equivalent moderate-intensity physical activity). The adjusted HR for CVD mortality was 0.57 (95% CI: 0.34−0.97) among participants who were insufficiently active and 0.69 (95% CI: 0.43−1.12) among those who were physically active. Among adults who were insufficiently active, the adjusted HR for all-cause mortality was 44% lower if engaging in muscle-strengthening activity ≥ 2 times/week. Thus, the increased opportunities for a healthy lifestyle in these countries that are associated with optimal climate conditions increase the opportunities for an active lifestyle during leisure time. In fact, walking and running (two of the most popular active ways of performing physical activity) are both fostered by climate conditions. A recent study that examined the relationship between leisure-time running and cardiovascular mortality risk [10] and that included a mean follow-up of 15 years of the participants, reported 3413 all-cause and 1217 cardiovascular deaths. It revealed that when runners were compared with nonrunners, they had 30% and 45% lower adjusted risks of all-cause and cardiovascular mortality, respectively, with a 3-year life expectancy benefit. Accordingly, as running has become very popular and is probably the easiest, cheapest and most accessible and enjoyable activity, it can be considered as a palliative to a sedentary lifestyle and a way to increase mental well-being in the general population. Clearly, when climate conditions are appropriate, the opportunities for running and an active lifestyle are enhanced.

Additionally, the association between physical activity during leisure time and active/sedentary lifestyles with regard to the incidence of mental disorders has been confirmed. Data from the NHANES I study found that people who reported little or no physical activity in their leisure time were more likely to exhibit more symptoms of depression [11]. These data were also confirmed in a dynamic prospective cohort of university graduates from a southern European country that were followed up for 6 years (the SUN cohort study, Seguimiento Universidad de Navarra in Spanish). In this study, we found a joint association of leisure-time physical activity and sedentary behavior on the incidence of mental disorders with those subjects with a physical activity level above the median showing a relative risk reduction of suffering a mental disorder of approximately 25% irrespective of their reported level on the sedentary index [odds ratio (OR) : 0.76; 95% CI : 0.61, 0.95 for those with a high sedentary index and OR : 0.75; 95% CI : 0.60, 0.93 for those with a low sedentary index] [12].

10.5   MENTAL WELL-BEING: THE IMPORTANCE OF SOCIALIZATION

As well as the positive effect of an active lifestyle fostered by the climate conditions, a central factor that contributes to this *"Mediterranean lifestyle"* is mental well-being. Mental well-being is associated with reduced levels of stress and results in a low incidence of depression and decreased suicide rates in the majority of the European Mediterranean countries.

In relation to stress levels, despite the physiological response to stress being well characterized, their possible links to cardiovascular risk are not totally understood. However, epidemiological data show that chronic stress predicts the occurrence of coronary heart disease. Moreover, a triggering factor for cardiac events among those individuals with advanced atherosclerosis is short-term emotional stress. Thus, a better understanding of stress as a cardiovascular risk factor and the use of stress management is now included in the European guidelines for CVD prevention [13].

As mentioned before, the psychological well-being of communities can also be affected by the climate and weather conditions. Multiple evidence exists suggesting that weather can have relevant effects on mood and mental well-being. One of the largest studies published in 1974 involving 16,000 students from Switzerland revealed that approximately 18% of the men and 29% of the women responded negatively to certain weather conditions, exhibiting symptoms of fatigue, dysphoric moods, irritability, and headaches [14]. Moreover, other studies have shown that humidity, temperature, and hours of sunshine had the greatest effect on people's mood [15].

Consequently, positive psychological aspects of well-being such as low stress levels and climate conditions are increasingly considered to play a protective role for cardiovascular disease and longevity [16]. A growing body of literature has linked positive well-being to better cardiovascular health, a lower incidence of CVD in healthy populations, and a reduced risk of adverse outcomes in patients with existing CVD.

It is also important to note that subjective well-being as measured through overall life satisfaction is very much shaped by socio-demographic factors such as age, income, or education which lead to different living situations as well as to different expectations and preferences. According to the 2015 Eurostat report "Quality of life in Europe—facts and views" that includes recent statistics on the quality of life of European Union citizens, unemployment is associated with very low life-satisfaction and at the same time that life satisfaction is clearly associated with income and education [17]. Therefore, these factors should also be taken into account when mental well-being is considered.

Sleep hygiene is another important factor that is also linked to better mental well-being. Several recent meta-analyses have shown that a U-shaped curve describes the relationship between sleep duration and CVD or all-cause mortality, with both short and long sleep durations being significant predictors of worse cardiovascular outcomes and higher all-cause mortality in prospective studies [18,19]. Another recent systematic review concluded that although a limited number of studies failed to demonstrate the association of subjective and objective sleep duration as well as sleep quality with noninvasive markers of subclinical CVD, most of the included studies showed a significant relationship between CVD and sleep hygiene [20]. In fact, it has been recently shown that poor sleep quality was highly prevalent and associated with depression and anxiety in cardiovascular patients [21].

A typical sleep pattern characteristic of some Mediterranean countries, the nap (also known as "siesta"), has also been studied in relation to CVD. A nap is a short sleep, classically taken during daylight hours. Daytime naps are usually brief, but can range from several minutes to several hours. In this regard, Yamada et al. [22] concluded that nap time and cardiovascular disease may be associated via a J-curve relation. In their dose-response meta-analysis, the authors concluded that a long daytime nap ($\geq$60 min/day) was associated with a higher risk of cardiovascular disease compared with not napping. On the contrary napping for <60 min/day was not associated with cardiovascular disease ($P = 0.98$) or all-cause mortality ($P = 0.08$), suggesting that further studies are needed to confirm the efficacy of a short nap.

Lastly, a central feature that influences people's mental well-being is the degree of socialization in their communities. Food and drink play a fundamental role in the holidays, fairs, and other celebrations that pack the public calendar, creating and sustaining social interaction and rebuilding a sense of community. Conviviality among citizens is of great importance for the general population's mental well-being. Many of the social interactions among individuals occur around the local food environment and in the parks and green areas. In this regard, many epidemiological studies have found that people living in environments with more green space report better physical and mental health than those with less green space [23].

Furthermore, in relation to the social role of the communities, what is of special relevance in people's perception of life satisfaction is supportive social relationships. In Europe, the life satisfaction degree is more than double (48.8% vs 19.0%) when people feel that "help is available when needed" compared to those who believe that "no help is available when needed" [17].

In conclusion, reduced stress levels, positive climate conditions, adequate sleep hygiene, an active lifestyle, and a noteworthy degree of conviviality and group socialization among others are included in and characterize the *"Mediterranean Lifestyle."* The fact that low depression and suicide rates are present in the Mediterranean countries is to a certain degree related to the general well-being of the population.

10.6 FUTURE PERSPECTIVES: URBAN ENVIRONMENTS AND THEIR RELATIONSHIP TO FOOD CONSUMPTION AND AN ACTIVE LIFESTYLE

There is a strong need to collaborate among researchers, practitioners, and residents to increase their mutual understanding of environmental influences on dietary behaviors [24] and lifestyles. Where you live and work may have a major impact on your health, as shown by previous research [25] and could be very different among countries and cultures. Urban environments may present a great opportunity for preventive population approaches [26]. Neighborhoods that are characterized as more walkable, either leisure-oriented or destination-driven, are associated with increased physical activity, increased social capital, lower overweight, lower reports of depression, and less-reported alcohol abuse [27].

The relationship between social and physical urban environment and cardiovascular health is an area of increasing interest that is currently under study among others in a Mediterranean country in the Heart Healthy Hoods (HHH) project, funded by the European Research Council (see www.hhhproject.eu). The combination of urban environmental measurements (both quantitative and qualitative) and universal electronic health records from the primary care health system is used in order to provide useful data that examine the relationship of neighborhood characteristics and cardiovascular health [28]. The HHH project focuses on four urban domains: physical activity, food, alcohol, and tobacco environments (see Fig. 10.1).

The study characterizes physical activity and food urban environments using a GIS-based multicomponent proposal [29] and assesses walking and cycling environments in the streets of the cities by comparing on-field and virtual audits [30], aiming to prevent noncommunicable diseases through

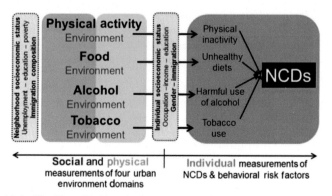

FIGURE 10.1   The Heart Healthy Hoods (HHH) study addresses the relationship between social and physical urban environment, individual behavioral risk factors, and noncommunicable diseases [31].

structural changes in urban environments [26]. In the future, comparison of urban environments among different cities, countries and geographical areas might be of help in order to better understand how the *"Mediterranean lifestyle"* can also be recognized applying this type of epidemiological tools.

REFERENCES

[1]   WHO. World Health Statistics 2016: Monitoring health for the sustainable development goals. Paris: World Health Organization; 2016.

[2]   Martinez-Gonzalez MA, Martin-Calvo N. Mediterranean diet and life expectancy; beyond olive oil, fruits, and vegetables. Curr Opin Clin Nutr Metab Care 2016;19(6):401−7.

[3]   Mackay J, Mensah GA. The atlas of heart disease and stroke. Brighton: World Health Organization; 2004.

[4]   OECD/EU. Health at a glance: Europe 2016—State of health in the EU cycle. Paris: OECD Publishing; 2016.

[5]   Cermakova P, Johnell K, Fastbom J, Garcia-Ptacek S, Lund LH, Winblad B, et al. Cardiovascular diseases in ∼30,000 patients in the Swedish dementia registry. J Alzheimers Dis 2015;48(4):949−58.

[6]   Kottek M, Rubel F. World maps of Köppen-Geiger Climate Classification. Vienna, Austria; 2017. Available from: <http://koeppen-geiger.vu-wien.ac.at/>.

[7]   UNWTO. UNWTO tourism highlights 2016 Edition. Madrid: World Tourism Organization; 2016.

[8]   Geleijnse JM. Vitamin D and the prevention of hypertension and cardiovascular diseases: a review of the current evidence. Am J Hypertens 2011;24(3):253−62.

[9]   Zhao G, Li C, Ford ES, Fulton JE, Carlson SA, Okoro CA, et al. Leisure-time aerobic physical activity, muscle-strengthening activity and mortality risks among US adults: the NHANES linked mortality study. Br J Sports Med 2014;48(3):244−9.

[10] Lee DC, Pate RR, Lavie CJ, Sui X, Church TS, Blair SN. Leisure-time running reduces all-cause and cardiovascular mortality risk. J Am Coll Cardiol 2014;64(5):472−81.

[11] Farmer ME, Locke BZ, Moscicki EK, Dannenberg AL, Larson DB, Radloff LS. Physical activity and depressive symptoms: the NHANES I Epidemiologic Follow-up Study. Am J Epidemiol 1988;128(6):1340−51.

[12] Sanchez-Villegas A, Ara I, Guillen-Grima F, Bes-Rastrollo M, Varo-Cenarruzabeitia JJ, Martinez-Gonzalez MA. Physical activity, sedentary index, and mental disorders in the SUN cohort study. Med Sci Sports Exerc 2008;40(5):827−34.

[13] Steptoe A, Kivimaki M. Stress and cardiovascular disease. Nat Rev Cardiol 2012;9 (6):360−70.

[14] Faust V, Weidmann M, Wehner W. The influence of meteorological factors on children and youths: a 10 per cent random selection of 16000 pupils and apprentices of Basle City (Switzerland). Acta Paedopsychiatr 1974;40(4):150−6.

[15] Howarth E, Hoffman MS. A multidimensional approach to the relationship between mood and weather. Br J Psychol 1984;75(Pt 1):15−23.

[16] Sin NL. The protective role of positive well-being in cardiovascular disease: review of current evidence, mechanisms, and clinical implications. Curr Cardiol Rep 2016;18 (11):106.

[17] Eurostat. Quality of life: Facts and views. Luxembourg: European Union; 2015.

[18] Cappuccio FP, D'Elia L, Strazzullo P, Miller MA. Sleep duration and all-cause mortality: a systematic review and meta-analysis of prospective studies. Sleep 2010;33(5):585−92.

[19] Cappuccio FP, Cooper D, D'Elia L, Strazzullo P, Miller MA. Sleep duration predicts cardiovascular outcomes: a systematic review and meta-analysis of prospective studies. Eur Heart J 2011;32(12):1484−92.

[20] Aziz M, Ali SS, Das S, Younus A, Malik R, Latif MA, et al. Association of subjective and objective sleep duration as well as sleep quality with non-invasive markers of subclinical cardiovascular disease (CVD): a systematic review. J Atheroscler Thromb 2017;24:208−26.

[21] Matsuda R, Kohno T, Kohsaka S, Fukuoka R, Maekawa Y, Sano M, et al. The prevalence of poor sleep quality and its association with depression and anxiety scores in patients admitted for cardiovascular disease: a cross-sectional designed study. Int J Cardiol 2017;228:977−82.

[22] Yamada T, Hara K, Shojima N, Yamauchi T, Kadowaki T. Daytime napping and the risk of cardiovascular disease and all-cause mortality: a prospective study and dose-response meta-analysis. Sleep 2015;38(12):1945−53.

[23] van den Berg M, van Poppel M, van Kamp I, Andrusaityte S, Balseviciene B, Cirach M, et al. Visiting green space is associated with mental health and vitality: a cross-sectional study in four European cities. Health Place 2016;38:8−15.

[24] Diez J, Conde P, Sandin M, Urtasun M, Lopez R, Carrero JL, et al. Understanding the local food environment: a participatory photovoice project in a low-income area in Madrid, Spain. Health Place 2016;43:95−103.

[25] Diez Roux AV, Mair C. Neighborhoods and health. Ann N Y Acad Sci 2010;1186:125−45.

[26] Franco M, Bilal U, Diez-Roux AV. Preventing non-communicable diseases through structural changes in urban environments. J Epidemiol Community Health 2015;69(6):509−11.

[27] Renalds A, Smith TH, Hale PJ. A systematic review of built environment and health. Fam Community Health 2010;33(1):68−78.

[28] Bilal U, Diez J, Alfayate S, Gullon P, Del Cura I, Escobar F, et al. Population cardiovascular health and urban environments: the Heart Healthy Hoods exploratory study in Madrid, Spain. BMC Med Res Methodol 2016;16:104.

[29] Cebrecos A, Diez J, Gullon P, Bilal U, Franco M, Escobar F. Characterizing physical activity and food urban environments: a GIS-based multicomponent proposal. Int J Health Geogr 2016;15(1):35.

[30] Gullon P, Badland HM, Alfayate S, Bilal U, Escobar F, Cebrecos A, et al. Assessing walking and cycling environments in the streets of Madrid: comparing on-field and virtual audits. J Urban Health 2015;92(5):923−39.

[31] Carreño V, Franco M, Gullón P, Carreño V. Studying city life, improving population health. Int J Epidemiol 2017;46(1):14−21 http://dx.doi.org/10.1093/ije/dyv207. PMID: 26604218

Chapter 11

A Healthy Diet for Your Heart and Your Brain

Almudena Sánchez-Villegas and Elena H. Martínez-Lapiscina

11.1 MEDITERRANEAN DIET AND DEMENTIA

11.1.1 Dementia: A Public Health Priority

According to the last update from the World Health Organization (WHO), 47.5 million people have dementia worldwide and there are 7.7 million new cases every year. More importantly, the total number of people living with dementia is expected to reach 75.6 million in 2030 and almost triple by 2050 to 135.5 million [1].

Despite the remarkable research efforts in order to achieve an effective drug to treat this condition; there is not any therapy currently available to cure dementia or even at least, dramatically halt the progression of the clinical course [1]. The sharp increase in the rates of this condition together with the absence of a cure explains that dementia is one of the major causes of disability and dependency among older people worldwide. Subsequently, the impact on families and caregivers as well as the social and economic consequences are challenging our economic capacity to attaining care for the growing burden of dementia.

In that light, WHO recognized dementia as a public health priority in the WHO report "Dementia: a public health priority," published in 2012 [2]. The goals of this report were to raise awareness about dementia and to commit some pioneering groups and institutions to lead and coordinate efforts on dementia. Beside other approaches focus on drug development, the efforts to halt rates of dementia thorough preventive strategies are essential as the WHO highlighted few months ago [1]. Although this last report recognized the relatively scarceness of research identifying modifiable risk factors of dementia, WHO recommended the prevention of vascular risk factors and the promotion of a healthy lifestyle as appropriate strategies to reduce risk of dementia [1]. The big impact of preventive strategies may be primarily explained by the fact that half of the cases of Alzheimer's disease, the

The Prevention of Cardiovascular Disease through The Mediterranean Diet.
DOI: http://dx.doi.org/10.1016/B978-0-12-811259-5.00011-1
169

leading cause of dementia worldwide, are attributable to seven potentially modifiable risk factors including well-known vascular risk factors [3]. If Alzheimer's disease could be delayed by even one single year, nine million cases would be eschewed over 40 years [3].

11.1.2 Dementia Is Incurable but Preventable: Relevance of Pathogenic Mechanisms

The most common causes of dementia include Alzheimer's disease, vascular dementia, and mixed dementia. Even though the ultimate etiological determinant of Alzheimer's disease remains unknown, it is believed that the accumulation of amyloid-beta plaques and tau-protein tangles are among the main factors leading neurodegeneration in the brain [4,5]. The etiology of vascular dementia, the second leading cause of dementia, involves vascular damage due to vascular risk factors that finally produce ischemia of strategic or/and extensive brain regions responsible for critical cognitive function [6]. Finally, mixed dementia is a condition where changes representing more than one type of dementia occur simultaneously in the brain. The most common pattern of mixed dementia is the presence of the plaques and tangles associated with Alzheimer's disease along with blood vessel changes associated with vascular dementia. There are studies that suggested that half of the brains of subjects diagnosed as having dementia had brain pathology compatible with mixed dementia [7]. Consequently, mixed dementia might be an under-recognized condition worldwide.

Despite the first etiological cause is far to be fully understood in the mentioned disorders, inflammation, oxidative stress, and vascular comorbidities are common pathogenic mechanisms acting as catalysts for spreading neurodegeneration in brain [8].

Oxidative stress represents a lack of balance between the production of reactive oxygen species and the autologous mechanisms capable to readily remove these reactive intermediates or to repair the subsequent damage. Neurons are prone to oxidative stress in as much as they need high requirements of oxygen consumption, they are inappropriately prepared with antioxidant defense system and have scarce capacity to damage repair [9]. There is increasing evidence of the relevant role of oxidative stress in Alzheimer's disease as well as in vascular dementia [10]. The presence of amyloid-beta plaques in Alzheimer's disease and the hypo perfusion in vascular dementia may produce mitochondrial dysfunction leading to increased oxidative stress that finally amplifies neurodegeneration in both diseases.

Inflammation is another relevant mechanism responsible for instigating disease progression in Alzheimer's disease. In this disorder, the accumulation of amyloid-beta plaques and tau-protein tangles foster microglial activation that promotes release of proinflammatory mediators as well as an increment in oxidative stress. This neuroinflammatory state induces increased amyloid deposition and neuroaxonal damage [11]. Similarly, the increased levels of proinflammatory mediators such as interleukin-6 (IL-6) and interleukin-18 (IL-18) in subjects harboring a diagnosis of vascular dementia supports a potential role of neuroinflammation in the pathophysiology of this disorder [12,13].

Finally, vascular comorbidities such as hypertension, dyslipidemia, diabetes mellitus, and obesity are major risk factors for cerebrovascular disease and therefore, vascular dementia [14]. In patients with Alzheimer's disease, vascular comorbidities cause microstructural vascular damage that challenges brain perfusion. As a result, hypoperfusion induce energy failure that negatively impacts the capacity of beta-amyloid clearance in the brain [15].

Combating the aforementioned mechanisms namely inflammation, oxidative stress, and vascular comorbidities may reduce the progression of pathogenic changes ultimately responsible for the appearance of dementia due to Alzheimer's or vascular disease. The potentially relevant impact of this approach on reducing rate of dementia may be explained in light of the concept of biological redundancy. Biological redundancy implies that the dysfunction of some cells in a biological system does not significantly impact the overall function of the system in as much as there are a myriad of cells responsible for same function in a system. For instance, it is well recognized that the onset of type 1 diabetes mellitus or Parkinson's disease barely appears before at least 90% of the beta cells of the pancreatic islets [16] or 50%−80% of the dopaminergic neurons of the substantia nigra are irreversibly damaged [17], respectively. These two examples clearly illustrate the relevance of maintaining the steady ongoing neurodegeneration below a threshold above which the clinical consequences due to pathological changes are prone to appear. Therefore, reducing the rate of development of pathological changes by counteracting the mentioned mechanisms does neither cure Alzheimer's nor vascular disease; however, it may be likely one of the most feasible strategies to reduce the appearance of dementia, the ultimate cause of patient's burden.

11.1.3 Dementia: Hope Through Mediterranean Diet

Beside other approaches to cope with inflammation, oxidative stress, and vascular comorbidities, the protective role of nutrition has been suggested in the last few decades. A huge number of studies have evaluated the ability of single food and nutrients in order to prevent cognitive decline or dementia but overall the results are inconclusive for supporting the recommendation in favor to any single food or nutrient in isolation [18]. The approach of evaluating the role of a single nutrient has been successful for understanding deficiency diseases, particularly, those related to the lack of vitamins.

However, the role of nutrition in chronic conditions may be seldom explained in light of single nutrients. By opposite, the harmonic combination of many of these nutrients and foods with potential capacity to reduce inflammation, oxidative stress, and promote vascular protection may likely overwhelm the hypothetical benefit of any single nutrient or food item [19]. Consequently, nutritional epidemiology shifted its focus to dietary patterns few years ago. A dietary pattern is defined as the quantity, variety, or combination of different foods and beverage in a diet and the frequency with which they are habitually consumed. There are some studies that support the potential benefit of different dietary patterns such as DASH (Dietary Approaches to Stop Hypertension) [20], Healthy Diet Index [21] and caloric restriction [22] on cognitive performance and/or dementia. However, the Mediterranean Diet has emerged as the most important dietary pattern capable to reduce cognitive decline and risk of dementia [23,24].

Other chapters of this book have deeply addressed antioxidative [25] and antiinflammatory [26] properties of Mediterranean Diet as well the favorable effect on vascular profile [27]. The benefits of the Mediterranean Diet support the recommendation of this dietary pattern as a feasible nonpharmacological strategy to combat the increased rate of dementia worldwide.

11.1.4 Epidemiological Evidences Regarding the Role of Mediterranean Diet in Cognitive Decline and Dementia

In 2006, Nicholas Scarmeas and colleagues published the first evidence of a protective role of the Mediterranean dietary pattern in the incidence of cognitive events in a large prospective cohort study. These authors described that participant from The Washington Heights/Hamilton Heights Aging Project (WHICAP) cohorts (inception years 1992 and 1999) with the highest adherence to a Mediterranean Diet displayed a 40% reduction in the risk of development Alzheimer's disease as to compared to those with an intermediate or low adherence to this dietary pattern [28]. Since then, an increased number of studies have evaluated the potential benefit of Mediterranean Diet in neurodegeneration and cognitive performance. Table 11.1 summarizes the main evidences regarding the adherence to the Mediterranean dietary pattern and the effect on cognitive performance and incidence of cognitive decline or dementia. The wide variability in the methodology of cognitive assessment challenges the comparability of the magnitude of the effect of the Mediterranean Diet on cognitive events. Nonetheless, the bulk of data suggest that a high adherence to a Mediterranean Diet may promote 18%−66% reduction in the risk of incidence of cognitive decline [29−31] and 34%−54% reduction in the risk of Alzheimer's disease [32−35].

Overall, the number of studies assessing the effect of the Mediterranean Diet on the rate of cognitive decline doubles the one addressing same effect on the incidence of mild cognitive impairment, Alzheimer's disease and all types of dementia. More importantly, the level of consistency achieved from observational studies supporting the benefit of Mediterranean Diet on cognitive decline [20,29,31,36−41] is conspicuously greater than the one from the

TABLE 11.1 Main Evidences Regarding the Association Between Adherence to the Mediterranean Dietary Pattern and Cognitive Performance and Incidence of Cognitive Decline or Dementia

Author, Year	Study, Country	Demographic Features (Sample Size; Sex; Age at First Dietary Assessment)	Cognitive Assessment	Results
Randomized Clinical Trials				
Knight, 2015	MedLey, Australia	166 ♂♀; age1 ≥ 65 years	Cognitive decline after 6 months of follow-up	Ongoing study
Hardman, 2015	LIILAC, Australia	148 ♂♀; age 60–90 years	Cognitive decline after 6 months of follow-up	Ongoing study
Unpublished	PREDIMED-PLUS, Spain	6,000 ♂♀; age ♂55–75 /♀60–75	Cognitive decline and incidence of cognitive events	Ongoing study
Valls-Pedret, 2017 (submitted)	PREDIMED, Spain	7,447 ♂♀; age 67 years	Incidence of Alzheimer's disease after 6 years of nutritional intervention	Benefit in favor MedDiet + Nuts: HR = 0.58; 95% CI (0.34–0.99)
Valls-Pedret, 2013	PREDIMED-Barcelona, Spain	334♂♀; age 67 years	Cognitive decline after 4 years of nutritional intervention	Significant reduction in rate of decline for MedDiet + EVOO and MedDiet + Nuts
Martinez-Lapiscina, 2013	PREDIMED-Navarra, Spain	557♂♀; age 67 years	Cognitive function after 6.5 years of nutritional intervention	Significant better performance for MedDiet + EVOO and MedDiet + Nuts

(Continued)

TABLE 11.1 (Continued)

Author, Year	Study, Country	Demographic Features (Sample Size; Sex; Age at First Dietary Assessment)	Cognitive Assessment	Results
Martinez-Lapiscina, 2013	PREDIMED-Navarra, Spain	268♂♀; age 67 years	Cognitive function and prevalence of MCI after 6.5 years of nutritional intervention	Significant better performance and lower prevalence of mild cognitive impairment for MedDiet + EVOO
McMillan, 2011	Australia	25♀; age 21 years	Cognitive decline after 10 days of nutritional intervention	Inconsistent results
Prospective Cohort Studies				
Haring, 2016	WHIMS, US	6425♀; age 65–79 years	Incidence of mild cognitive impairment or probable dementia after 9.1 years of follow-up	No association
Qin, 2015	CHNS, China	1650♂♀; age 63–64 years	Cognitive decline after 5.3 years of follow-up	Significant benefit in favor MedDiet
Gardener, 2015	AIBL, Australia	527♂♀; age 69.3 years	Cognitive decline after 3 years of follow-up	Significant benefit in favor MedDiet
Trichopoulou, 2015	EPIC, Greece	401♂♀; age ≥ 60 years	Cognitive decline after 6–8 years of follow-up	Benefit in favor MedDiet: substantial decline vs stable condition, high vs other adherence OR = 0.34; 95% CI (0.13–0.89)

Study	Cohort	Sample	Outcome	Results
Morris, 2015	MAP, US	923♂♀; age 58–98 years	Incidence of Alzheimer's disease after 4.5 years of follow-up	Benefit in favor MedDiet: highest vs lowest adherence HR = 0.46; 95% CI (0.26–0.79)
Galbete, 2015	SUN, Spain	823♂♀; age 61.9 years	Cognitive decline after 2 years of follow-up	Significant benefit in favor MedDiet
Koyama, 2014	Health ABC, US	2326♂♀; age 74.6 years	Cognitive decline after 7.9 years of follow-up	Significant benefit in favor MedDiet (only for black ethnicity)
Tangney, 2014	MAP, US	826♂♀; age 81.5 years	Cognitive decline after mean follow-up of 4.1 years	Significant benefit in favor MedDiet
Tsivgoulis, 2013	REGARDS, US	17,478♂♀; age 64.6 years	Incidence of cognitive impairment after 4 years of follow-up	Benefit in favor MedDiet: highest vs lowest adherence OR = 0.87; 95% CI (0.76–1.00)
Samieri, 2013	NHS, US	16,058♀; age 61.3–74.1 years	Cognitive decline after 6 years of follow-up	Significant benefit in favor MedDiet: highest adherence to MedDiet equivalent to delaying cognitive aging by one year
Samieri, 2013	WHS, US	6174♀; age 66.3 years	Cognitive decline after 6 years of follow-up	No association
Titova, 2013	PIVUS, Sweden	194♂♀; age 70 years	Cognitive function after 5 years of dietary assessment	No association
Wengreen, 2013	CCMS, US	3831♂♀; age 74.1 years	Cognitive decline after 11 years of follow-up	Significant benefit in favor MedDiet

(Continued)

TABLE 11.1 (Continued)

Author, Year	Study, Country	Demographic Features (Sample Size; Sex; Age at First Dietary Assessment)	Cognitive Assessment	Results
Kesse-Guyot, 2013	SU.VI.MAX, France	3083♂♀; age 52 years	Cognitive function after 13 years of dietary assessment	Significant benefit in favor MedDiet but some inconsistency due to different MedDiet scores used
Vercambre, 2012	WACS, US	2504 ♀; age 68.4–69.1 years	Cognitive decline after 5.4 years of follow-up	No association
Tangney, 2011	CHAP, US	3790♂♀; age 75.4 years	Cognitive decline after 7.6 years of follow-up	Significant benefit in favor MedDiet
Cherbuin, 2011	PATH-TLS, Australia	1528♂♀; age 62.5 years	Incidence of mild cognitive impairment (different definitions) after 4 (original) and 8 (abstract) years of follow-up	No association
Roberts, 2010	MCSA, US	1141♂♀; age 79.7–83.3 years	Incidence of mild cognitive impairment or dementia after 2.2 years of follow-up	No significative association: highest vs lowest MedDiet HR = 0.75; 95% CI (0.46–1.21)
Gu, 2010	WHICAP 1992 and 1999, US	1219♂♀; age 76.7 years	Incidence of Alzheimer's disease	Benefit in favor MedDiet: highest vs lowest adherence HR = 0.66; 95% CI (0.41–1.04)
Féart, 2009	3CS, France	1410♂♀; age 75.9 years	Cognitive decline and incidence of dementia and Alzheimer's disease after 4.1 years of follow-up	Significant benefit in favor MedDiet for one cognitive test, not for others and not for dementia and Alzheimer's disease

Scarmeas, 2009	WHICAP 1992 and 1999, US	1880♂♀; age 77.2 years	Incidence of Alzheimer's disease after 5.4 years of follow-up	Benefit in favor MedDiet: highest vs lowest adherence HR = 0.60; 95% CI (0.42–0.87)
Scarmeas, 2009	WHICAP 1992 and 1999, US	1393♂♀; age 76.7 years for healthy population 482♂♀; age 77.5 years for population with mild cognitive impairment	Incidence of mild cognitive impairment and progression from mild cognitive impairment to Alzheimer's disease after 4.5 years of follow-up	Benefit in favor MedDiet: highest vs lowest adherence HR = 0.72; 95% CI (0.52–1.00) for mild cognitive impairment and HR = 0.52; 95% CI (0.30–0.91) for transition from cognitively impaired to Alzheimer's disease
Psaltopoulou, 2008	EPIC, Greece	732♂♀; age ≥ 60 years	Cognitive function after 8 years of dietary assessment	Nonsignificant week associations in favor MedDiet
Scarmeas, 2006	WHICAP 1992 and 1999, US	2258♂♀; age 77.2 years	Incidence of Alzheimer's disease after 4 years of follow-up	Benefit in favor MedDiet: highest vs lowest adherence HR = 0.60; 95% CI (0.42–0.87)

MedLey, The Mediterranean diet for cognitive function and cardio-vascular health in the elderly; *LILAC Study*, the Lifestyle Intervention in Independent Living Aged Care; *PREDIMED-PLUS*, PREvencion con DIeta MEDiterranea PLUS; *MedDiet*, Mediterranean Diet; *EVOO*, Extra-Virgin Olive Oil; *PREDIMED*, PREvencion con DIeta MEDiterranea; *WHIMS*, Women's Health Initiative Memory Study; *CHNS*, China Health and Nutrition Survey; *AIBL*, Australian Imaging, Biomarkers and Lifestyle study of ageing; *EPIC*, European Prospective Investigation into Cancer and Nutrition; *MAP*, Rush Memory and Aging Project; *SUN*, Seguimiento Universidad de Navarra; *Health ABC*, Health, Aging and Body Composition study; *REGARDS*, Reasons for Geographic and Racial Differences in Stroke; *NHS*, Nurses' Health Study; *WHS*, Women's Health Study; *PIVUS*, Prospective Investigation of the Vasculature in Uppsala Seniors; *CCMS*, Cache County Memory Study; *SU.VI.MAX*, The Supplementation with Vitamins and Mineral Antioxidants study; *WACS*, Women's Antioxidant Cardiovascular Study; *CHAP*, Chicago Health and Aging Project; *PATH-TLS*, PATH Through Life study; *MCSA*, Mayo Clinic Study of Aging; *WHICAP*, Washington/Hamilton Heights-Inwood Columbia Aging Project; *3CS*, French Three City cohort.

studies assessing the same effect on the incidence of hard events such as mild cognitive impairment or dementia [30,33−35] (Table 11.1). These findings suggest that the benefit in rate of cognitive decline may be methodologically easier to address than the benefit in the incidence of cognitive impairment or dementia. Besides methodological issues, biological aspects may be taken into consideration in order to further explain these findings. Mild cognitive impairment and dementia typically appear at the latest phases of neurodegenerative conditions such as Alzheimer's or vascular degenerative disease. According to the aforementioned explanations, dementia or notorious cognitive impairment emerges when biological redundancy is exceeded. The studies focused on evaluating the incidence of these events may be likely including population with substantial burden of neuroaxonal injury, close to or even overwhelming the threshold above which biological redundancy may not maintain an appropriate cognitive function. Even though one study has shown benefit in this situation [30], combating inflammation, oxidative stress, and vascular comorbidities at the end of the neurodegenerative processes may have limited effect in order to improve the clinical course of the disease.

In line with aforementioned facet, another critical factor highlighted in the literature is the age of the participants in observational studies. Some authors have focused their research efforts on elderly populations. By selecting this target population, they try to increase the efficiency of their designs because the majority of the potential events of cognitive decline would be expected to appear in elderly participants. However, this strategy may be appropriate only if they successfully evidence that they have quantified adherence to the Mediterranean Diet pattern in the middle-aged of participants and if so, that the level of adherence has been maintained over the follow-up in the cohort study. To sum-up, the best strategy would be firstly recruiting a middle-aged population; secondly, monitoring the compliance to recommendations of the Mediterranean Diet in order to evaluate the level of consistency in the adherence to the pattern among participants and finally, measuring cognitive decline after several years of follow-up. This strategy allows evaluating the long-term effect of the Mediterranean Diet in a middle-aged population, for whom the benefits are expected to be the highest in order to halt the risk of incidence of cognitive decline, a condition that typically appears in the senescence. Since this strategy is methodologically complex and particularly expensive, most studies have evaluated the effect of the Mediterranean Diet on participants with ages comprised between 65 and 75 years at the age of the first dietary assessment. Among the studies targeting younger participants (age <65 years) [29,40−42], a sole study has found negative results [43]. This study including participants from the Australian, the PATH Through Life Study, did not find any significant association between the adherence to Mediterranean Diet and the risk of cognitive decline or the transition from normal aging to mild cognitive impairment. However, some aspects regarding the source of some key nutrients and the cooking style of some food might at least partially explain these negative results [43].

This Australian study introduces another relevant aspect to take into consideration in order to interpret result from studies outside the Mediterranean basis. The Mediterranean Diet has become so widespread that some countries have included modifications in the traditional Mediterranean pattern in order to align recommendations with cultural tastes in countries outside Mediterranean basis. The aim of this strategy in order to promote a quick shift to a healthy Mediterranean dietary pattern worldwide is praiseworthy because this approach tries to spread benefits of the Mediterranean Diet throughout the world to reach epidemic proportions of people living in a healthy lifestyle. However, these modifications should be considered with caution in as much as some of these changes involve key elements of the pattern such as olive oil and Mediterranean cooking styles such as baking or grilling. In line with this, the mentioned Australian study found that the high consumption of fish was associated with increased risk of cognitive impairment. However, the authors recognized that the cooking style might be behind this odd association. The fish is likely more frequently consumed as "fish and chips" than as "grilled fish served with vegetables" in Australia. One could not expect to that fried fish and chips may lead similar benefits that grilled fish served with vegetables. Moreover, the authors found that a high intake of monounsaturated fats was predictive of mild cognitive impairment. Similarly, the authors recognized that the origin of these fatty acids in this study might be trans fats instead of olive oil [43]. Likewise, cultural differences in relation with the olive oil consumption may explain lack of significant results in the Three Cities Cohort study [44]. In this French cohort, the authors found a protective effect of Mediterranean Diet in reducing rate of cognitive decline in some test but did not find any significant results for dementia or Alzheimer's disease. In this study, the consumption of olive oil was scored as intensive by less than 40% of participants. More importantly, a single weekly use of olive oil for cooking or dressing was the requirement to be considered intense [44]. Participants from the Greek EPIC (European Prospective Investigation into Cancer and Nutrition) cohort consumed about 46–52 g/day [45]; similarly, participants from the PREvención con DIeta MEDiterránea in Spanish (PREDIMED) trial reported daily consumptions of 40 g before enrollment [46]. The whole is more important than the sum of its parts [19] but reducing the consumption of olive oil may be comparable to renounce to key player in any team. Consequently, changes promoting aligning with recommendations of the Mediterranean Diet may be desirable to widespread the "magic pill" as long as these changes do not undermine the benefits of the Mediterranean dietary pattern.

There are a couple of randomized clinical trials that have evaluated the effects of Mediterranean Diet on cognitive decline. The first trial included 25 young (21 years) females who were assigned to a 10 days of dietary change or to maintain the usual diet. The study has a significant number of methodological issues that may explain the inconsistency of the results [47]. The strongest evidence in favor of Mediterranean Diet as a recommendation for the brain healthy has come from PREDIMED Study. The results from three studies run in two different nodes of PREDIMED (Navarra and Barcelona) were consistent and found that participants assigned to a Mediterranean Diet supplemented with either extra-virgin olive oil or nuts displayed better cognitive performance [48,49] and lower cognitive decline [50] as well as lower prevalence of mild cognitive impairment (only for the group assigned to extra-virgin olive oil supplementation) [49] at the end of 6.5 years of a nutritional intervention promoting Mediterranean Diet as to compared to a low-fat diet. Finally, recent analyses including the entire PREDIMED study population ($n = 7447$) observed a reduced risk of the incidence of Alzheimer's disease (101 events) for participants allocated to the group of Mediterranean Diet supplemented with nuts [hazard ratio (HR) : 0.58; 95% confidence interval (CI) : 0.34−0.99] compared to the group assigned to a low-fat diet (Valls-Pedret C, submitted, 2017).

11.1.5   Ongoing Efforts: From Dietary Toward Holistic Interventions

Overall, the previous studies supported the role of promoting Mediterranean Diet in order to prevent neurodegeneration in healthy subjects. Nevertheless, evidence from new randomized clinical trials is advisable in order to strengthen this recommendation. Fortunately, these are some ongoing studies including two Australian randomized trials [51,52] and the PREDIMED-PLUS trial in Spain [53]. The two Australian trials are primarily designed to address the effect of an intervention promoting Mediterranean Diet on short-term cognitive decline (6 months) compared to a standard diet [51,52]. The MedLey study will evaluate the effect of the Mediterranean Diet in isolation [52] whereas the LIILAC study will compare three interventional arms, namely, dietary intervention (Mediterranean Diet), exercise intervention (aerobic physical activity), and a combined dietary and exercise intervention to the standard dietary pattern [51]. The PREDIMED-PLUS trial, a natural evolution of PREDIMED trial, will use a similar approach in order to further the evidence supporting the role of a healthy lifestyle to counteract burden due to chronic conditions, particularly cardiovascular disorders. The PREDIMED-PLUS trial is aimed to demonstrate that the benefit from a holistic intervention promoting healthy lifestyle including the energy-restricted Mediterranean pattern, regular and low-intense physical activity (45 minutes of walking or equivalent) and a mild weight loss (8%)

overwhelms the benefit attained from the dietary intervention with Mediterranean Diet in isolation. Once again, the whole is more important than the sum of its parts; likewise a healthy lifestyle may have a deeper impact on health than diet in isolation. The main idea behind these studies is to promote a healthy lifestyle as a feasible nonpharmacological strategy to counteract the challenging sharp rise of the burden due to chronic conditions including neurodegenerative disorders.

Overall, the results available so far support the role of the Mediterranean Diet as an approach to reduce cognitive decline. Mediterranean Diet may be especially successful if the adherence is promoted during the middle age and maintained for long period of time.

11.1.6   Recommendations

At the International Conference on Nutrition and the Brain, Washington, DC, July 19−20, 2013, speakers were asked to comment on possible guidelines for Alzheimer's disease prevention, with an aim of developing a set of practical steps to be recommended to members of the public. From this discussion, seven guidelines emerged related to healthful diet and exercise habits [54].

1. Minimize your intake of saturated fats and trans fats. Saturated fat is found primarily in dairy products, meats, and certain oils (coconut and palm oils). Trans fats are found in many snack pastries and fried foods and are listed on labels as "partially hydrogenated oils."
2. Vegetables, legumes (beans, peas, and lentils), fruits, and whole grains should replace meats and dairy products as primary staples of the diet.
3. Vitamin E should come from foods, rather than supplements. Healthful food sources of vitamin E include seeds, nuts, green leafy vegetables, and whole grains. The recommended dietary allowance (RDA) for vitamin E is 15 mg per day.
4. A reliable source of vitamin B12, such as fortified foods or a supplement providing at least the recommended daily allowance (2.4 mg/day for adults), should be part of your daily diet. Have your blood levels of vitamin B12 checked regularly as many factors, including age, may impair absorption.
5. If using multiple vitamins, choose those without iron and copper and consume iron supplements only when directed by your physician.
6. Although aluminum's role in Alzheimer's disease remains a matter of investigation, those who desire to minimize their exposure can avoid the use of cookware, antacids, baking powder, or other products that contain aluminum.
7. Include aerobic exercise in your routine, equivalent to 40 minutes of brisk walking three times per week.

However, most of the evidence until now comes from observational studies prone to confusion especially in lifestyle-related strategies. Future well-designed randomized clinical trials will specifically evaluate the benefits of diet and specifically Mediterranean Diet on neurodegeneration and hopefully; the results may strengthen the magnitude of the recommendation. In the meantime, the certain benefit of the Mediterranean Diet to prevent cardiovascular conditions should be taken into consideration and on this basis, adherence to the Mediterranean dietary pattern should be advisable.

11.2   MEDITERRANEAN DIET AND DEPRESSION

11.2.1   Diet, a New Target to Prevent Depression

Unipolar disorder is one of the most common mental disorders which represents a condition seriously recurrent associated with worse quality of life, elevated health and pharmacologic costs, important personal and familiar suffering, frequent comorbidities, and an increment in suicide risk and total mortality. Unipolar depression affects more than 151 million people worldwide, is the worldwide the second leading cause of years of healthy life lost as a result of disability [55] and it is projected to also be the leading cause of disability-adjusted life years lost in 2030 [56].

Classically, one of the objectives of the nutritional epidemiology has been to analyze the role of diet in the prevention of some noncommunicable diseases such as cardiovascular disease and little attention has been paid to the effect on other diseases such as mental disorders [57]. However, an emerging field of research is currently being developed and several links between nutrition and mental health are being established. There are a number of scientifically rigorous studies making important contributions to the understanding of the role of nutrition in mental health [58]. Specifically, some studies have pointed out that several dietary patterns could be associated with a reduced prevalence [59] and a reduced risk of depression among adults [60,61]. This association seems to be consistent across countries, cultures, and populations according to several systematic reviews and meta-analyses [62,63].

11.2.2   Etiological Hypothesis of Depression

Major depressive disorder is considered to be a multifactorial disease. Therefore, a wide array of potential causes including biological, psychological, and environmental factors are likely to be involved in the etiology of depression.

Well-defined clinical and laboratory evidence supports a hormonal role in depression. This hypothesis holds that repeated or maintained stress-induced hypercortisolemia will eventually lead to downregulation of glucocorticoid receptors in the Central Nervous System. Chronic stress activates the hypothalamus-pituitary-adrenal axis (HPA-axis) and the sympathetic nervous system, increasing the production of cortisol in the adrenal cortex. The normal negative feedback on cortisol consisting in inhibition of corticotropin-releasing hormone and adrenocorticotrophic hormone as a response to high cortisol levels will be eventually reversed in depression. As a consequence, an alteration of the HPA-axis is frequently seen in depressed patients with positive instead of negative feedback. This creates a vicious cycle with cortisol levels permanently elevated. The consequence of this vicious circle is that levels of cortisol, corticotropin-releasing hormone, and adrenocorticotrophic hormone are often simultaneously elevated in depressed patients. This vicious circle has also been supposedly implicated in reduced volume of the hippocampus, reduced rates of neurogenesis, accumulation of visceral fat, abdominal obesity, a higher propensity to insulin resistance, and an increase in the production of inflammatory cytokines [64].

Another widely accepted etiological theory is the monoamine hypothesis of depression. It suggests that depression is related to a dysfunction in serotoninergic or noradrenergic systems in the Central Nervous System. In this context, an adequate intake of vitamin B, folate, and vitamin B12 can be speculated to be associated with lower depression risk because they are involved in the metabolism of S-adenosyl methionine and methionine (an essential amino acid). These latter two compounds are critical to the production of monoamines (serotonin and noradrenaline) and methylation in the brain. An interesting hypothesis of depression suggests that deficiencies in vitamin B6, folate, and vitamin B12 can lead to reduced synthesis of these monoamines and decreased availability of them in the synapse. The repeatedly reported association between elevated levels of homocysteine and depression seems to corroborate this hypothesis because suboptimal levels of folate and B vitamins are related with higher homocysteine levels and higher homocysteine levels have been associated with depression through several mechanisms such as its detrimental vascular effect or the excitotoxic effect of some of its metabolites such as homocysteinic acid or cysteine sulfinic acid [65].

A large number of studies have demonstrated that depressive disorders are accompanied by the activation of several inflammatory mechanisms that lead to the increment in some pro-inflammatory cytokines such as IL-1b, IL-2, and IL-6, or umor necrosis factor-α (TNF-α). In fact, several studies have established the possible role of inflammation in depression through several mechanisms such as activation of the HPA-axis. Inflammation also causes and up-regulation of indoleamine 2,3-dioxygenase (a rate limiting enzyme in the metabolism of tryptophan), that leads to neurotoxicity and serotonin deficiency in the brain, which are core features of the pathophysiology of depression. Finally, high levels of proinflammatory cytokines are also responsible for alteration in neurotransmitters metabolism and decrease in brain-derived neurotrophic factor (BDNF) availability. BDNF is a peptide that is critical for axonal growth, neuronal survival, and synaptic plasticity.

Finally, it has been suggested that perturbations of cerebral endothelium may mediate progressive neuronal dysfunction. Endothelial cells synthesize and secrete BDNF [65]. Some studies as well as a recent meta-analysis have found a relationship between endothelial function and the presence of depression [66].

11.2.3   Depression, Cardiovascular Risk Factors, and Cardiovascular Disease

Cardiovascular disease and depression are the two single leading causes of disability worldwide and they are frequently comorbid conditions with an estimated prevalence of comorbidity of around 20%. Depression shares common mechanisms with obesity, metabolic syndrome, type 2 diabetes, and cardiovascular disease. In fact, the comorbidity of depression and cardiovascular risk factors is frequent [67]. Metabolic and inflammatory processes, such as reduced insulin sensitivity, elevations in plasma homocysteine levels and more importantly, increased production of proinflammatory cytokines and endothelial dysfunction, seem to be the major factors responsible for the link between depression and cardio-metabolic disorders.

A bidirectional relationship between depression and cardiovascular disease has been proposed to explain the frequent co-occurrence of both conditions. A possible explanation for the high comorbidity of depression and cardiovascular disease is a common pathophysiology. The neurobiological results in this field support a central role of a systemic proinflammatory state as a common pathophysiological mechanism in both conditions, with a shared profile of inflammation biomarkers, including neutrophil gelatinase-associated lipocalin (NGAL) as a promising biomarker linking cardiovascular disease and depression [68], along with C-reactive protein or IL-6. Inflammation is associated with endothelial dysfunction and atherosclerotic progression (and hence with cardiovascular disease).

Regarding the comorbidity between type 2 diabetes and depression, in 2015, two different reviews indicated that both diseases might have a common etiology, diabetes increasing the prevalence or risk for future depression; depression increasing the prevalence or risk for future diabetes [69,70].

A key candidate for a common pathway could be the activation and disturbance of the stress system and chronic hypercortisolemia that promote insulin resistance, visceral obesity, and lead to metabolic syndrome and diabetes. In fact, in the Whitehall II cohort study, low insulin secretion was associated with an increased risk of developing depressive symptoms [71].

Moreover, chronic stress and induces immune dysfunction. High amounts of inflammatory cytokines interact with the normal functioning of the pancreatic β-cells, induce insulin resistance, and thus, promote the appearance of diabetes. In several studies, insulin resistance has been also linked to endothelial dysfunction or inflammation processes. Indeed, not only metabolic disturbances but also inflammatory markers have been associated with depressive symptoms in participants with diabetes from the SEARCH for Diabetes in Youth cohort study [72]. Thus, stress and inflammation both would promote depression and diabetes, giving a feasible common link between them.

11.2.4 Biological Plausibility for the Link Between Cardio-Protective Mediterranean Diet and Reduced Risk of Depression

The cardio-protective effect of the Mediterranean Diet is, as we have deeply mentioned, due mainly to its antiinflammatory and antioxidant properties as well as its capacity to improve endothelial function, insulin sensitivity, and glycemic control or reduce the risk of several cardiovascular risk factors such as obesity, metabolic syndrome, or diabetes.

All these conditions are common links between cardiovascular disease and depression as it has been mentioned above. Beyond the improvement of cardiovascular risk, the Mediterranean Diet has potential for enhanced synaptogenesis and neurogenesis as a consequence of better availability and function of neurotrophins. Moreover, this diet ensures nutritional adequacy and availability of B vitamins and folate related to the synthesis of monoamines. Green leafy vegetables, legumes, grains, nuts, and fruits have a high nutritional value and are important sources of minerals and folic acid. Finally, its content in fish and nuts improves lipid composition (omega-3 fatty acids and other unsaturated fatty acids instead of trans-fatty acids or saturated fatty acids) of cell membranes where receptors for neurotransmitters are located.

Table 11.2 summarizes the main effects of the Mediterranean Diet that could explain its beneficial role on depression.

11.2.5 Epidemiological Evidences Regarding the Role of Mediterranean Diet in Depression

Although the scientific report of the 2015 American dietary guidelines advisory committee has concluded that current evidence is limited, the protective dietary patterns associated with reduced risk of depression are those patterns emphasizing seafood, vegetables, fruits, and nuts [73]. These features would correspond with the definition of several cardio-protective dietary patterns including the

TABLE 11. 2 Main Effects of the Mediterranean Diet That Could Explain Its Beneficial Role on Depression

Nutritional adequacy: bio-availability of B vitamins and minerals (related to the synthesis of monoamines)

Improvement of lipid subtypes (omega-3 fatty acids and other unsaturated fatty acids instead of trans-fatty acids or saturated fatty acids) in cell membranes

Antiinflammatory properties

Reduction of oxidative stress (antioxidant effect that produces neuroprotection)

Improvement of endothelial function (increase of neurotrophins synthesis, including brain-derived neurotrophic factor)

Improvement in glycemic control (insulin sensitivity, and leptin and glucose levels)

Reduction in the risk of diabetes, metabolic syndrome and cardiovascular disease

Weight control

Mediterranean Diet. Table 11.3 summarizes the main evidences associating the adherence to several dietary patterns and depression.

In fact, age-adjusted rates of depression and suicide are known to be lower in Southern Europe (Spain, Italy, and Greece) than in Northern or Central European countries. This South-to-North gradient has been hypothetically related to the diversity of food patterns between Mediterranean and non-Mediterranean countries. This difference also indirectly suggests that some aspects of the traditional Mediterranean Diet might have contributed to lower the risk of depression.

The role of Mediterranean Diet in depression was first evaluated in 2009 within the SUN (Seguimiento Universidad de Navarra) Project. The SUN Project is a cohort study based on university graduates from Spain who respond to different questionnaires every two years. These questionnaires collect information regarding dietary habits, lifestyle, the use of medication, and presence of diseases, among other variables. A total of 10,094 initially healthy Spanish participants free of depression at baseline were followed up for a median of 4 years. A reduced risk of depression associated with closer conformity to the traditional Mediterranean Diet was found. The conformity to the Mediterranean Diet was operationally defined using the Mediterranean Diet Score (MDS) proposed by Trichopoulou et al., which ranges from 0 to 9 points. Adherence to the MDS was categorized into five categories: low (score 0−2), low moderate (score 3), moderate-high (score 4), high (score 5), and very high (6−9). Taking the category of lowest adherence as reference the researchers found relative risk reductions for the 4 upper successive categories of adherence to the MDS (26%, 34%, 51%, and 42%, respectively) with a significant dose−response relationship [74].

TABLE 11. 3 Main Evidences Associating the Adherence to Several Dietary Patterns and Depression

Author, Year	Study, Country	Pattern	Outcome	Results
Randomized Clinical Trials (RCTs)				
Sanchez-Villegas, 2013	PREDIMED, Spain	MedDiet	Medical diagnosis	No significant reduction
Observational Cohort Studies				
Sanchez-Villegas, 2009	SUN, Spain	MedDiet	Medical diagnosis	Significant reduction
Skarupski, 2013	CHAP, US	MedDiet	CES-D	Significant reduction
Psaltopoulou, 2008	EPIC, Greece	MedDiet	GDS	No significant association
Sanchez-Villegas, 2015	SUN, Spain	MedDiet	Medical diagnosis	Significant reduction
Sanchez-Villegas, 2015	SUN, Spain	PDP	Medical diagnosis	Significant reduction
Sanchez-Villegas, 2015	SUN, Spain	AHEI	Medical diagnosis	Significant reduction
Akbaraly, 2013	WHITEHALL II, UK	AHEI	CES-D	Reduction in women, nonsignificant in men
Shatenstein, 2012	NuAge, Canada	HEI	GDS	No significant association

(Continued)

TABLE 11.3 (Continued)

Author, Year	Study, Country	Pattern	Outcome	Results
Lai, 2015	ALSWH, Australia	ARFS	CES-D	Significant reduction with baseline data for intake
Lai, 2016	ALSWH, Australia	ARFS	CES-D	No significant association with updated data for intake
Rienks, 2013	ALSWH, Australia	PC analysis - MedDiet	CES-D	Significant reduction
Akbaraly, 2009	WHITEHALL II, UK	PC analysis	CES-D	Traditional significant reduction Processed significant increment
Ruusunen, 2014	KIHD, Finland	PC analysis	Medical diagnosis	Prudent significant reduction
Jacka, 2014	PTHS, Australia	PC analysis	Goldberg DS	Prudent significant reduction Western significant increment (attenuation by socioec. and health variables)
Le Port, 2012	GAZEL Cohort, France	PC analysis	CES-D	Western, fat-sweet, snack significant increment Healthy significant reduction

Chan, 2014	Hong Kong, China	PC analysis	GDS	No significant association
Chocano-Bedoya, 2013	NHS, US	PC analysis	Medical diagnosis	No significant association
Vermeulen, 2016	InChianti study, Italy	RRR-MedDiet	CES-D	Significant reduction
Lucas, 2014	NHS, US	RRR-Inflammatory index	Medical diagnosis	Significant increment
Sanchez-Villegas, 2015	SUN, Spain	DII	Medical diagnosis	Significant increment

PREDIMED, PREvencion con DIeta MEDiterranea; *SUN*, Seguimiento Universidad de Navarra; *NuAge*, Québec Longitudinal Study on Nutrition and Aging; *ALSWH*, Australian Longitudinal Study on Women's Health; *KIHD*, Kuopio Ischemic Heart Disease Risk Factor Study; *PTHS*, Personality and Total Health Study; *CHAP*, Chicago Health and Aging Project; *EPIC*, European Prospective Investigation into Cancer and Nutrition; *NHS*, Nurses' Health Study; *MedDiet*, Mediterranean Diet; *PDP*, Pro-vegetarian Dietary Pattern; *AHEI*, Alternative Healthy Eating Index; *HEI*, Healthy Eating Index; *ARFS*, Australian Recommended Food Score; *PC*, Principal Component; *RRR*, Reduced rank regression; *DII*, Dietary Inflammatory Index; *CES-D*, Center for Epidemiologic Studies Depression Scale; *GDS*, Geriatric Depression Scale; *Goldberg DS*, Goldberg Depression Scale.

Recently, these results have been confirmed in an updated analysis including repeated measures of Mediterranean Diet adherence [75]. Other observational longitudinal studies, have also found a protective effect of this pattern on depression development. Several of them were carried out in Mediterranean countries. It is the case of a recent analysis in the Chianti cohort study, a study based on 1362 participants aged 18−102 years. The adherence to a dietary pattern rich in vegetables, olive oil, grains, fruit, fish, and moderate in wine and red and processed meat that the authors labeled as "typical Tuscan dietary pattern" and that could be defined such as typical Italian Mediterranean dietary pattern was associated with lower depressive symptoms over a 9-year period of follow-up [76]. A total of 732 men and women, 60 years or older, participating in the Greek EPIC cohort were also analyzed in relation to risk factors for cognitive function and depressive symptoms in the elderly. The adherence to the Mediterranean Diet as well as olive oil consumption and monounsaturated fatty acids intake (most important fatty acid contained in olive oil) were associated to better scores in both conditions although the results were not statistically significant [77].

On the other hand, some epidemiological studies have been carried out in other non-Mediterranean populations. Rienks et al. analyzed the association between the adherence to the Mediterranean Diet and depression among a sample of mid-aged women from the Australian Longitudinal Study on Women's Health. A higher consumption of the Mediterranean-style diet had a cross-sectional association with lower prevalence of depressive symptoms and longitudinally with lower incidence of depressive symptoms (risk reduction: 17%) [78].

In the Chicago Health and Aging Project, greater adherence to a Mediterranean-based diet was associated with a reduced number of newly occurring depressive symptoms over an average follow-up of 7.2 years. The annual rate of developing depressive symptoms was 98.6% lower among persons in the highest tertile of a Mediterranean dietary pattern compared with persons in the lowest tertile group [79].

Nevertheless, the available evidence is still sparse and not definitive. Moreover, interpretation of findings resulting from observational studies demand caution. Most of the studies had a cross-sectional design, which is a weak design for inferring cause and effect relationships that can only be suggested. In such studies, exposure is ascertained simultaneously with disease and, therefore, an alternative interpretation of the results could be made as a consequence of reverse causation bias; for example, that depression may lead to poorer dietary habits. Moreover, one of the most important limitations in observational epidemiology is to obtain adequate control of confounding. This is a key methodological issue because potential effects of dietary patterns on depression could be explained in part by the cooccurrence of other lifestyle-related and sociodemographic factors, or by medical conditions closely related to the adherence to a particular dietary pattern.

This issue could be solved by carrying out large randomized trials with interventions based on changes in the overall food pattern. Currently, several intervention trials have started up to assess the role of Mediterranean Diet in depression but at this moment, none of them has any published result. The SMILES study is an intervention trial with healthy diet based on Australian and Greek dietary guidelines. The dietary support group demonstrated, in recent results, significantly greater improvement in depressive symptoms and higher rate of remission than the social support control group [80]. However, the aim of the trial is to ascertain the effectiveness of the dietary intervention to treat acute depression (the diet would serve as a coadjuvant) not to prevent it [80].

On the other hand, the MedLey study is designed to specifically determine the effect of an intervention with a Mediterranean dietary pattern in elder subjects but considering as primary outcome their cognitive function and not a clinical diagnosis of depression although the design includes to assess aspects related to psychological well-being like stress, anxiety, sleeping patterns, or depression [81].

The MooDFOOD prevention trial is a multicountry intervention trial that examines the feasibility and effectiveness of two different nutritional strategies (multinutrient supplementation and food-related behavioral change therapy) to prevent depression in individuals who are overweight and have elevated depressive symptoms but who are not currently or in the last 6 months meeting criteria for an episode of major depressive disorder [82].

Only the PREDIMED study has tested the effects on the risk of incident cases of clinical depression during 5.4-year follow-up. The point estimates for the relative risks (RR) associated with the intervention using a Mediterranean Diet supplemented with extra-virgin olive oil (RR = 0.91, 9% risk reduction) or a Mediterranean Diet supplemented with mixed tree nuts (RR = 0.78, 22% risk reduction) in the PREDIMED trial suggested an inverse association. However, the confidence intervals for both estimates were wide and they showed that the results were compatible with a null result. Even when both Mediterranean diets were merged together and analyzed as a single group, the results were not statistically significant. Only when the analysis was limited to participants with type 2 diabetes (approximately 50% of the sample), a significantly reduced risk of depression was observed and only for participants assigned to the Mediterranean Diet + nuts (significant risk reduction: 41%) but not for participants randomly allocated to Mediterranean Diet + extra-virgin olive oil [83].

11.2.6   Recommendations

Although there are a number of important gaps in the scientific literature to date, existing evidence suggests that a combination of healthful dietary practices (including the adherence to the Mediterranean Diet) may reduce the risk of developing depression. In 2015, a group of experts with several backgrounds (epidemiology, psychiatry, and nutrition) elaborated the first Dietary

Recommendations for the Prevention of Depression [84]. These dietary recommendations also provided additional and/or concurrent benefits for obesity, cardiovascular disease, diabetes, and metabolic syndrome, and essentially pose no risk of harm.

1. Follow "traditional" Mediterranean Diet. The available evidence suggests that this traditional dietary pattern may be beneficial for positive mental health.
2. Increase your consumption of fruits, vegetables, legumes, whole grain cereals, nuts and seeds. These foods should form the bulk of the diet as they are nutrient dense, high in fiber and low in saturated and trans fatty acids.
3. Include a high consumption of foods rich in omega-3 polyunsaturated fatty acids (PUFAs). Fish is one of the main sources of omega-3 PUFAs, and higher fish consumption is associated with reduced depression risk.
4. Limit your intake of processed-foods, "fast" foods, commercial bakery goods, and sweets. These foods are high in trans fatty acids, saturated fat, refined carbohydrates, and added sugars, and are low in nutrients and fiber. Consumption of these foods has been associated with an increased risk or probability of depression in observational studies.
5. Replace unhealthy foods with wholesome nutritious foods. The Mediterranean dietary pattern (e.g., fruits, vegetables, whole-grain cereals, and fish or olive oil) and unhealthy dietary patterns (e.g., sweets, soft-drinks, fried food, refined cereals, and processed meats) are independent predictors of lower and higher depressive symptoms, respectively.

As the body of evidence grows from controlled intervention studies on dietary patterns and depression, these recommendations should be modified accordingly.

REFERENCES

[1] World Health Organization (WHO). Available from: <http://www.who.int/mediacentre/factsheets/fs362/en/>.
[2] World Health Organization (WHO). Available from: <http://www.who.int/mental_health/publications/dementia_report_2012/en/>.
[3] Barnes DE, Yaffe K. The projected effect of risk factor reduction on Alzheimer's disease prevalence. Lancet Neurol 2011;10(9):819–28.
[4] Hardy J, Allsop D. Amyloid deposition as the central event in the aetiology of Alzheimer's disease. Trends Pharmacol Sci 1991;12(10):383–8.
[5] Goedert M, Spillantini MG, Crowther RA. Tau proteins and neurofibrillary degeneration. Brain Pathol 1991;1(4):279–86.
[6] Roman GC, Tatemichi TK, Erkinjuntti T, Cummings JL, Masdeu JC, Garcia JH, et al. Vascular dementia: diagnostic criteria for research studies. Report of the NINDS-AIREN international workshop. Neurology 1993;43(2):250–60.
[7] Schneider JA, Arvanitakis Z, Bang W, Bennett DA. Mixed brain pathologies account for most dementia cases in community-dwelling older persons. Neurology 2007;69(24):2197–204.

[8] Vijayan M, Reddy PH. Stroke, vascular dementia, and Alzheimer's disease: molecular links. J Alzheimers Dis 2016;54(2):427–43.

[9] Halliwell B. Oxidative stress and neurodegeneration: Where are we now? J Neurochem 2006;97(6):1634–58.

[10] Luca M, Luca A, Calandra C. The role of oxidative damage in the pathogenesis and progression of Alzheimer's disease and vascular dementia. Oxid Med Cell Longev 2015;2015:504678.

[11] Calsolaro V, Edison P. Neuroinflammation in Alzheimer's disease: current evidence and future directions. Alzheimers Dement 2016;12(6):719–32.

[12] Malaguarnera L, Motta M, Di Rosa M, Anzaldi M, Malaguarnera M. Interleukin-18 and transforming growth factor-beta 1 plasma levels in Alzheimer's disease and vascular dementia. Neuropathology 2006;26(4):307–12.

[13] Miwa K, Okazaki S, Sakaguchi M, Mochizuki H, Kitagawa K. Interleukin-6, interleukin-6 receptor gene variant, small-vessel disease and incident dementia. Eur J Neurol 2016;23 (3):656–63.

[14] Frances A, Sandra O, Lucy U. Vascular cognitive impairment, a cardiovascular complication. World J Psychiatry 2016;6(2):199–207.

[15] de la Torre JC. Cardiovascular risk factors promote brain hypoperfusion leading to cognitive decline and dementia. Cardiovasc Psychiatry Neurol 2012;2012:367516.

[16] American Diabetes Association. Diagnosis and classification of diabetes mellitus. Diabetes care 2009;32(Suppl 1)):S62–7.

[17] DeMaagd G, Philip A. Parkinson's disease and its management: Part 1: Disease entity, risk factors, pathophysiology, clinical presentation, and diagnosis. PT 2015;40(8):504–32.

[18] Williams JW, Plassman BL, Burke J, Benjamin S. Preventing Alzheimer's disease and cognitive decline. Evid Rep Technol Assess (Full Rep) 2010;193:1–727.

[19] Martinez-Gonzalez MA, Martin-Calvo N. Mediterranean diet and life expectancy; beyond olive oil, fruits, and vegetables. Curr Opin Clin Nutr Metab Care 2016;19(6):401–7.

[20] Wengreen H, Munger RG, Cutler A, Quach A, Bowles A, Corcoran C, et al. Prospective study of dietary approaches to stop hypertension- and Mediterranean-style dietary patterns and age-related cognitive change: the cache county study on memory, health and aging. Am J Clin Nutr 2013;98(5):1263–71.

[21] Correa Leite ML, Nicolosi A, Cristina S, Hauser WA, Nappi G. Nutrition and cognitive deficit in the elderly: a population study. Eur J Clin Nutr 2001;55(12):1053–8.

[22] Le Bourg E. Dietary restriction studies in humans: focusing on obesity, forgetting longevity. Gerontology 2012;58(2):126–8.

[23] Cao L, Tan L, Wang HF, Jiang T, Zhu XC, Lu H, et al. Dietary patterns and risk of dementia: a systematic review and meta-analysis of cohort mtudies. Mol Neurobiol 2016;53(9):6144–54.

[24] Petersson SD, Philippou E. Mediterranean Diet, cognitive function, and dementia: a systematic review of the evidence. Adv Nutr 2016;7(5):889–904.

[25] Sureda A, Del Mar Bibiloni M, Martorell M, Buil-Cosiales P, Marti A, Pons A, et al. Mediterranean diets supplemented with virgin olive oil and nuts enhance plasmatic antioxidant capabilities and decrease xanthine oxidase activity in people with metabolic syndrome: the PREDIMED study. Mol Nutr Food Res 2016;60(12):2654–64.

[26] Casas R, Sacanella E, Urpi-Sarda M, Corella D, Castaner O, Lamuela-Raventos RM, et al. Long-term immunomodulatory effects of a Mediterranean Diet in adults at high risk of cardiovascular disease in the PREvencion con DIeta MEDiterranea (PREDIMED) randomized controlled trial. J Nutr 2016;146(9):1684–93.

[27] Ros E, Martinez-Gonzalez MA, Estruch R, Salas-Salvado J, Fito M, Martinez JA, et al. Mediterranean diet and cardiovascular health: teachings of the PREDIMED study. Adv Nutr 2014;5(3):330s−6s.

[28] Scarmeas N, Stern Y, Mayeux R, Luchsinger JA. Mediterranean diet, Alzheimer disease, and vascular mediation. Arch Neurology 2006;63(12):1709−17.

[29] Tsivgoulis G, Judd S, Letter AJ, Alexandrov AV, Howard G, Nahab F, et al. Adherence to a Mediterranean diet and risk of incident cognitive impairment. Neurology 2013;80 (18):1684−92.

[30] Scarmeas N, Stern Y, Mayeux R, Manly JJ, Schupf N, Luchsinger JA. Mediterranean diet and mild cognitive impairment. Arch Neurology 2009;66(2):216−25.

[31] Trichopoulou A, Kyrozis A, Rossi M, Katsoulis M, Trichopoulos D, La Vecchia C, et al. Mediterranean diet and cognitive decline over time in an elderly Mediterranean population. Eur J Nutr 2015;54(8):1311−21.

[32] Gu Y, Luchsinger JA, Stern Y, Scarmeas N. Mediterranean diet, inflammatory and metabolic biomarkers, and risk of Alzheimer's disease. J Alzheimers Dis 2010;22(2):483−92.

[33] Scarmeas N, Luchsinger JA, Schupf N, Brickman AM, Cosentino S, Tang MX, et al. Physical activity, diet, and risk of Alzheimer disease. JAMA 2009;302(6):627−37.

[34] Scarmeas N, Stern Y, Tang MX, Mayeux R, Luchsinger JA. Mediterranean diet and risk for Alzheimer's disease. Ann Neurol 2006;59(6):912−21.

[35] Morris MC, Tangney CC, Wang Y, Sacks FM, Bennett DA, Aggarwal NT. MIND diet associated with reduced incidence of Alzheimer's disease. Alzheimers Dement 2015;11(9): 1007−14.

[36] Tangney CC, Kwasny MJ, Li H, Wilson RS, Evans DA, Morris MC. Adherence to a Mediterranean-type dietary pattern and cognitive decline in a community population. Am J Clin Nutr 2011;93(3):601−7.

[37] Tangney CC, Li H, Wang Y, Barnes L, Schneider JA, Bennett DA, et al. Relation of DASH- and Mediterranean-like dietary patterns to cognitive decline in older persons. Neurology 2014;83(16):1410−16.

[38] Gardener SL, Rainey-Smith SR, Barnes MB, Sohrabi HR, Weinborn M, Lim YY, et al. Dietary patterns and cognitive decline in an Australian study of ageing. Mol Psychiatry 2015;20(7):860−6.

[39] Koyama A, Houston DK, Simonsick EM, Lee JS, Ayonayon HN, Shahar DR, et al. Association between the Mediterranean diet and cognitive decline in a biracial population. J Gerontol A Biol Sci Med Sci 2015;70(3):354−9.

[40] Qin B, Adair LS, Plassman BL, Batis C, Edwards LJ, Popkin BM, et al. Dietary patterns and cognitive decline among Chinese older adults. Epidemiology 2015;26(5):758−68.

[41] Galbete C, Toledo E, Toledo JB, Bes-Rastrollo M, Buil-Cosiales P, Marti A, et al. Mediterranean diet and cognitive function: the SUN project. J Nutr Health Aging 2015;19(3): 305−12.

[42] Kesse-Guyot E, Andreeva VA, Lassale C, Ferry M, Jeandel C, Hercberg S, et al. Mediterranean diet and cognitive function: a French study. Am J Clin Nutr 2013;97(2): 369−76.

[43] Cherbuin N, Anstey KJ. The Mediterranean diet is not related to cognitive change in a large prospective investigation: the PATH Through Life study. Am J Geriatr Psychiatry 2012;20(7):635−9.

[44] Feart C, Samieri C, Rondeau V, Amieva H, Portet F, Dartigues JF, et al. Adherence to a Mediterranean diet, cognitive decline, and risk of dementia. JAMA 2009;302(6):638−48.

[45] Psaltopoulou T, Kyrozis A, Stathopoulos P, Trichopoulos D, Vassilopoulos D, Trichopoulou A. Diet, physical activity and cognitive impairment among elders: the EPIC-Greece cohort (European Prospective Investigation into Cancer and Nutrition). Public Health Nutr 2008;11(10):1054−62.

[46] Zazpe I, Sánchez-Tainta A, Estruch R, Lamuela-Raventos RM, Schroder H, Salas-Salvado J, et al. A large randomized individual and group intervention conducted by registered dietitians increased adherence to Mediterranean-type diets: the PREDIMED study. J Am Diet Assoc 2008;108(7):1134−44 discussion 45.

[47] McMillan L, Owen L, Kras M, Scholey A. Behavioural effects of a 10-day Mediterranean diet. Results from a pilot study evaluating mood and cognitive performance. Appetite. 2011;56(1):143−7.

[48] Martinez-Lapiscina EH, Clavero P, Toledo E, Estruch R, Salas-Salvado J, San Julian B, et al. Mediterranean diet improves cognition: the PREDIMED-NAVARRA randomised trial. J Neurol Neurosurg Psychiatry 2013;84(12):1318−25.

[49] Martinez-Lapiscina EH, Clavero P, Toledo E, San Julian B, Sánchez-Tainta A, Corella D, et al. Virgin olive oil supplementation and long-term cognition: the PREDIMED-NAVARRA randomized, trial. J Nutr Health Aging 2013;17(6):544−52.

[50] Valls-Pedret C, Sala-Vila A, Serra-Mir M, Corella D, de la Torre R, Martinez-Gonzalez MA, et al. Mediterranean Diet and age-related cognitive decline: a randomized clinical trial. JAMA Intern Med 2015;175(7):1094−103.

[51] Hardman RJ, Kennedy G, Macpherson H, Scholey AB, Pipingas A. A randomised controlled trial investigating the effects of Mediterranean diet and aerobic exercise on cognition in cognitively healthy older people living independently within aged care facilities: the Lifestyle Intervention in Independent Living Aged Care (LIILAC) study protocol [ACTRN12614001133628]. Nutr J 2015;14:53.

[52] Knight A, Bryan J, Wilson C, Hodgson J, Murphy K. A randomised controlled intervention trial evaluating the efficacy of a Mediterranean dietary pattern on cognitive function and psychological wellbeing in healthy older adults: the MedLey study. BMC geriatrics 2015;15:55.

[53] PREDIMED-PLUS. Available from: <http://www.predimedplus.com/>.

[54] Neal D, Barnard ND, Bush AI, Ceccarelli A, Cooper J, de Jager CA, et al. Dietary and lifestyle guidelines for the prevention of Alzheimer's disease. Neurobiology of Aging 2014;35:S74−8.

[55] Murray CJ, Vos T, Lozano R, Naghavi M, Flaxman AD, Michaud C, et al. Disability-adjusted life years (DALYs) for 291 diseases and injuries in 21 regions, 1990−2010: a systematic analysis for the global burden of disease study 2010. Lancet 2013;380 (9859):2197−223.

[56] Global Burden of Disease Study 2013. Collaborators. Global, regional, and national incidence, prevalence, and years lived with disability for 301 acute and chronic diseases and injuries in 188 countries, 1990-2013: a systematic analysis for the Global Burden of Disease Study 2013. Lancet 2015;386(9995):743−800.

[57] Sanchez-Villegas A, Martínez-González MA. Diet, a new target to prevent depression? BMC Med 2013;11:3.

[58] Sarris J, Logan AC, Akbaraly TN, Amminger GP, Balanzá-Martínez V, Freeman MP, et al. Nutritional medicine as mainstream in psychiatry. Lancet Psychiatry 2015;2:271−4.

[59] Ruusunen A, Lehto SM, Mursu J, Tolmunen T, Tuomainen TP, Kauhanen J, et al. Dietary patterns are associated with the prevalence of elevated depressive symptoms and the risk

of getting a hospital discharge diagnosis of depression in middle-aged or older Finnish men. J Affect Disord 2014;159:1−6.

[60] Akbaraly TN, Sabia S, Shipley MJ, Batty GD, Kivimaki M. Adherence to healthy dietary guidelines and future depressive symptoms: evidence for sex differentials in the Whitehall II study. Am J Clin Nutr 2013;97(2):419−27.

[61] Jacka FN, Cherbuin N, Anstey KJ, Butterworth P. Dietary patterns and depressive symptoms over time: examining the relationships with socioeconomic position, health behaviours and cardiovascular risk. PLoS One 2014;9(1):e87657.

[62] Lai J, Hiles S, Bisquera A, Hure AJ, McEvoy M, Attia J. A systematic review and meta-analysis of dietary patterns and depression in community-dwelling adults. Am J Clin Nutr 2014;99:181−97.

[63] Rahe C, Unrath M, Berger K. Dietary patterns and the risk of depression in adults: a systematic review of observational studies. Eur J Nutr 2014;53(4):997−1013.

[64] Martínez-González MA, Sánchez-Villegas A. Food patterns and the prevention of depression. Proc Nutr Soc 2016;75(2):139−46.

[65] Guo S, Kim WJ, Lok J, Lee SR, Besancon E, Luo BH, et al. Neuroprotection via matrix-trophic coupling between cerebral endothelial cells and neurons. Proc Natl Acad Sci USA 2008;105(21):7582−7.

[66] Cooper DC, Tomfohr LM, Milic MS, Natarajan L, Bardwell WA, Ziegler MG, et al. Depressed mood and flow-mediated dilation: a systematic review and meta-analysis. Psychosom Med 2011;73:360−9.

[67] GBD 2013. Mortality and Causes of Death Collaborators. Global, regional, and national age-sex specific all-cause and cause-specific mortality for 240 causes of death, 1990-2013: a systematic analysis for the global burden of disease study 2013. Lancet 2015;385 (9963):117−71.

[68] Gouweleeuw L, Naude PJ, Rots M, DeJongste MJ, Eisel UL, Schoemaker RG. The role of neutrophil gelatinase associated lipocalin (NGAL) as biological constituent linking depression and cardiovascular disease. Brain Behav Immun 2015;46:23−32.

[69] Berge LI, Riise T. Comorbidity between type 2 diabetes and depression in the adult population: directions of the association and its possible pathophysiological mechanisms. Int J Endocrinol 2015;2015:164760.

[70] Moulton CD, Pickup JC, Ismail K. The link between depression and diabetes: the search for shared mechanisms. Lancet Diabetes Endocrinol 2015;3:461−71.

[71] Akbaraly TN1, Kumari M, Head J, Ritchie K, Ancelin ML, Tabák AG, et al. Glycemia, insulin resistance, insulin secretion, and risk of depressive symptoms in middle age. Diabetes Care 2013;36(4):928−34.

[72] Hood KK, Lawrence JM, Anderson A, Bell R, Dabelea D, Daniels S, Rodriguez B, et al. Metabolic and inflammatory links to depression in youth with diabetes. Diabetes Care 2012;35:2443−6.

[73] Dietary Guidelines Advisory Committee. Scientific Report of the 2015. Dietary Guidelines Advisory Committee, Washington, D.C.; 2015.

[74] Sánchez-Villegas A, Delgado-Rodríguez M, Alonso A, Schlatter J, Lahortiga F, Serra Majem L, et al. Association of the Mediterranean dietary pattern with the incidence of depression: the Seguimiento Universidad de Navarra/University of Navarra follow-up (SUN) cohort. Arch Gen Psychiatry 2009;66(10):1090−8.

[75] Sánchez-Villegas A, Henríquez-Sánchez P, Ruiz-Canela M, Lahortiga F, Molero P, Toledo E, et al. A longitudinal analysis of diet quality scores and the risk of incident depression in the SUN Project. BMC Med 2015;13:197.

[76] Vermeulen E, Stronks K, Visser M, Brouwer IA, Schene AH, Mocking RJ, et al. The association between dietary patterns derived by reduced rank regression and depressive symptoms over time: the Invecchiare in Chianti (InCHIANTI) study. Br J Nutr 2016;115(12): 2145−53.

[77] Psaltopoulou T, Kyrozis A, Stathopoulos P, Trichopoulos D, Vassilopoulos D, Trichopoulou A. Diet, physical activity and cognitive impairment among elders: the EPIC-Greece cohort (European Prospective Investigation into Cancer and Nutrition). Public Health Nutr 2008;11(10):1054−62.

[78] Rienks J, Dobson AJ, Mishra GD. Mediterranean dietary pattern and prevalence and incidence of depressive symptoms in mid-aged women: results from a large community-based prospective study. Eur J Clin Nutr 2013;67(1):75−82.

[79] Skarupski KA, Tangney CC, Li H, Evans DA, Morris MC. Mediterranean diet and depressive symptoms among older adults over time. J Nutr Health Aging 2013;17(5):441−5.

[80] Jacka FN, O'Neil A, Opie R, Itsiopoulos C, Cotton S, Mohebbi M, et al. A randomised controlled trial of dietary improvement for adults with major depression (the "SMILES" trial). BMC Med 2017;15(1):23.

[81] Knight A, Bryan J, Wilson C, Hodgson J, Murphy K. A randomised controlled intervention trial evaluating the efficacy of a Mediterranean dietary pattern on cognitive function and psychological wellbeing in healthy older adults: the MedLey study. BMC Geriatr. 2015;15:55.

[82] Roca M, Kohls E, Gili M, Watkins E, Owens M, Hegerl U, et al. MooDFOOD prevention trial investigators. prevention of depression through nutritional strategies in high-risk persons: rationale and design of the MooDFOOD prevention trial. BMC Psychiatry 2016;16:192.

[83] Sánchez-Villegas A, Martínez-González MA, Estruch R, Salas-Salvadó J, Corella D, Covas MI, et al. Mediterranean dietary pattern and depression: the PREDIMED randomized trial. BMC Med 2013;11:208.

[84] Opie RS, Itsiopoulos C, Parletta N, Sanchez-Villegas A, Akbaraly TN, Ruusunen A, et al. Dietary recommendations for the prevention of depression. Nutr Neurosci 2015;20(3): 161−71.

Chapter 12

The Mediterranean Cook: Recipes for All Seasons

María Soledad Hershey and Ana Sánchez-Tainta

Salads and Vegetables
- Orange and Raisin Endive Salad
- Feta Cheese and Nut Spinach Salad
- Cream of Butternut Squash
- Tomato Soup
- "Escalibada" Toast

Pasta and Rice
- Pasta With Red Peppers and Tuna
- Pea and Mushroom Spaghetti
- Whole-Grain Noodles With Mussels
- Vegetable Soup With Brown Rice
- Seafood Paella

Legumes
- Chickpea and Egg Salad
- Chickpeas With Codfish
- Mediterranean Lentils
- Greek "Fasolada"
- Sautéed Peas

Egg
- Spanish Omelet With "Pisto"
- Broccoli and Cheese Omelet
- Scrambled Eggs With Asparagus
- Poached Eggs With Clams and Marinara Sauce
- Fried Eggs With Mushrooms

Fish
- Sole With Clams
- Squid Skewers
- Codfish With Onion and Tomato Sauce

The Prevention of Cardiovascular Disease through The Mediterranean Diet.
DOI: http://dx.doi.org/10.1016/B978-0-12-811259-5.00012-3

- Grilled Salmon With Rice
- Swordfish With Tomato and Black Olives

Meats
- Baked Chicken With Vegetables
- Chicken Thighs With Olives
- Turkey Cooked With Tomatoes and Peppers
- Turkey With Sautéed Zucchini and Onion Sandwich
- Pork Tenderloin With Apple Sauce

12.1 SALADS AND VEGETABLES

12.1.1 Orange and Raisin Endive Salad

Ingredients *Yields 4 servings*

- 4 endives
- 1¼ cups of raisins
- 2 oranges
- ⅔ cup of yogurt
- Vinegar
- 2 tsp. of extra-virgin olive oil
- A pinch of salt

Directions

Chop the endives and place them on a platter. Peel the oranges and cut them into small pieces. Add the pieces of orange and raisins to the platter. Make the sauce with yogurt, olive oil, a drizzle of vinegar, and a pinch of salt. Pour the mixture over the salad and serve.

12.1.2 Feta Cheese and Nut Spinach Salad

Ingredients *Yields 4 servings*

- 1¼ cups of fresh spinach
- ¼ cup of unshelled walnuts
- ½ cup of feta cheese
- ½ cup of cherry tomatoes
- 3 Tbsp. of extra-virgin olive oil
- 4 tsp. of balsamic vinegar
- A pinch of salt

Directions

Rinse, drain, and place the spinach on a platter. Chop and add the walnuts with the cubed feta cheese to the platter. Cut the cherry tomatoes in half and add them to the salad. Make the vinaigrette with the olive oil, balsamic vinegar, and a pinch of salt. Toss the salad with the vinaigrette just before serving to prevent the spinach from becoming too soft.

12.1.3   Cream of Butternut Squash

Ingredients *Yields 4 servings*

- 2½ cups of butternut squash
- 1¼ cups of carrots, chopped
- 1¼ cups of leeks, chopped
- ½ cup of shredded cheese
- 4 tsp. of extra-virgin olive oil
- A pinch of salt

Directions

Wash and cut all the vegetables. Sauté them for a few minutes in a pot with olive oil and then cover them with water. Cook at low heat for 30 minutes. Afterward, blend in a blender and distribute into individual bowls. Top with cheese and bake in the oven until the tops are golden before serving.

12.1.4   Tomato Soup

Ingredients *Yields 4 servings*

- 3½ cups of plum tomatoes
- 1 onion
- 1 clove of garlic
- 1 tsp. of sugar
- 5 Tbsp. of extra-virgin olive oil
- Basil
- A pinch of salt

Directions

Wash and cut the tomatoes. Cut the onion in thin long strips and sauté with olive oil and a clove of garlic. Add the tomato and cook everything for a minute. Add the sugar and a pinch of salt and blend until obtaining a creamy consistency. If it is very thick, you can add half a cup of water. Store in the refrigerator for an hour and garnish with a little basil before serving.

12.1.5   "Escalibada" Toast

Ingredients *Yields 4 servings*

- 8 slices of bread
- 1 eggplant

- 1 red pepper
- 1 green pepper
- 2 tomatoes
- 1 onion
- 3 Tbsp. of extra-virgin olive oil
- A pinch of salt

Directions

Preheat the oven to 360°F. Place the whole vegetables with a pinch of salt and a drizzle of olive oil on an ovenproof platter. Bake in the oven for an hour and once cooled, peel the vegetables. Take out the seeds from the eggplant and peppers, cut them into strips, and mix with the juices that they released in the oven. Cut the tomatoes and the onion. Toast the slices of bread in the oven for 5 minutes and place the "escalibada" over each of the toasts. Add a drizzle of olive oil on top before serving.

12.2 PASTA AND RICE

12.2.1 Pasta With Red Peppers and Tuna

Ingredients *Yields 4 servings*

- 10 oz. of short pasta
- 2 roasted red peppers
- 2 cans of tuna in olive oil
- 1 clove of garlic
- 3 Tbsp. of extra-virgin olive oil
- A pinch of salt

Directions

Cook the pasta in boiling water with salt and a little olive oil for 10 minutes, drain, toss, and drain again. Cut the garlic clove into thin slices and cook until golden in a pan with 2 Tbsp. of olive oil and add the roasted red peppers. Add a pinch of salt and add to the pasta. Crumble the tuna and add to the pasta and peppers.

12.2.2 Pea and Mushroom Spaghetti

Ingredients *Yields 4 servings*

- 10 oz. of spaghetti
- 1 cup of mushrooms (7 oz.)
- 1 small can of peas
- 1¼ cups of tomato sauce
- 1 clove of garlic

- 4 Tbsp. of extra-virgin olive oil
- A pinch of salt

Directions

Cut the garlic into thin slices and sauté in a pan with olive oil. Once golden, add the washed and sliced mushrooms and cook for 10 minutes covered at low heat. Drain the peas and add them to the mushrooms. Mix this combination with the tomato sauce and cook for a few minutes. Cook the pasta in a pot of boiling water the time indicated on the package and once cooked, drain and mix with the prepared sauce.

12.2.3 Whole-Grain Noodles With Mussels

Ingredients *Yields 4 servings*

- 10 oz. of whole-grain noodles
- 3 cloves of garlic
- ½ lb. of mussels, cooked
- Extra-virgin olive oil
- Parsley
- Salt

Directions

Cook the whole-grain noodles in boiling water with salt for the time indicated on the package. Drain the pasta and set aside. Peel the cloves of garlic and cut into thin slices. In a pan with heated olive oil sauté the garlic without letting it burn. Add the cooked mussels and sauté everything together. Add the cooked pasta and stir so the ingredients mix homogenously. Sprinkle parsley on top and serve.

12.2.4 Vegetable Soup With Brown Rice

Ingredients *Yields 4 servings*

- A butternut squash
- 3.5 oz. of green beans
- 1 carrot
- 1 leek
- ½ cup of brown rice
- ¾ gallon of water
- Ground black pepper
- Extra-virgin olive oil
- Salt

Directions

In a pan with olive oil, cut the carrot and the leek into thin strips, the green beans into flat strips, and dice the butternut squash. Poach all together for 3 minutes while continually stirring. Immediately after, pour the vegetables into a pot of water and let them cook for 20 minutes. Add the rice and cook until the rice is cooked *al dente*, about 45 minutes. Add a pinch of salt and serve hot.

12.2.5 Seafood Paella

Ingredients *Yields 4 servings*

- 1¼ cups of round rice
- 1 cup of fresh calamari
- ½ cup of shrimp
- 12 mussels
- 2 cloves of garlic
- ½ cup of tomato sauce
- 2½ cups of fish broth
- Saffron or yellow food coloring
- A pinch of salt

Directions

Pour olive oil into the paella pan and once it is hot sauté the shrimp and mussels for a minute. Take them out and set them aside. Cut the calamari in squares. In the same paella pan sauté the calamari. Once they are golden, add the tomato sauce and minced garlic, and sauté the mixture. Next, add the rice, stir, and sauté the mixture. Afterward pour two times the amount of broth than rice. Add the saffron or, instead, a pinch of food coloring. Add salt and let the paella cook at low heat.

After 10 minutes of cooking, place the shrimp and mussels on top of the rice and let it cook for another 5 more minutes until the broth has evaporated completely.

Before serving, let it sit for 5 minutes away from the heat.

12.3 BEANS

12.3.1 Chickpea and Egg Salad

Ingredients *Yields 4 servings*

- 1½ cups of cooked chickpeas
- 1 egg
- 2 tomatoes

- ¼ cup of green pepper, chopped
- 4 Tbsp. of extra-virgin olive oil
- 4 tsp. of balsamic vinegar
- Parsley
- A pinch of salt

Directions

Wash and cut the tomato and green pepper into cubes. Add half of the olive oil and balsamic vinegar, and a pinch of salt to the cut vegetables. Let this base for the salad soften for 30 minutes. Cook the egg in boiling water with a little bit of salt for 10 minutes, let it cool, peel, and shred the hardboiled egg with a grater.

Mix the softened tomato with the cooked and drained chickpeas in a bowl. Top off the salad with olive oil and a touch of salt and sprinkle the shredded egg on top.

12.3.2 Chickpeas With Codfish

Ingredients *Yields 4 servings*

- 1½ cups of chickpeas
- 1 bunch of spinach
- 1 onion
- ½ lb. of dry codfish
- 2 cloves of garlic
- 1 dried red pepper
- Extra-virgin olive oil
- Salt

Directions

Defrost the codfish 24−35 hours before preparing the dish. Put the chickpeas to soak in cold water the day before. Drain the chickpeas from the soaking water and cook them in a pressure cooker for 15−20 minutes. Peel and chop the onion. Sauté the onion in a deep pot with olive oil. Add the dried red peppers and stir with a spoon. Add the spinach to the pot and sauté the mixture for 10 minutes. When the chickpeas have cooked, add them to the pot with the spinach. In a small pot with boiling water, dip the codfish in for 2 minutes and then let it cool. Remove the skin and bones, shred the meat, and add it to the stew half an hour before the stew is done cooking. Serve hot once it has sat for a few minutes.

12.3.3 Mediterranean Lentils

Ingredients *Yields 4 servings*

- 1 cup of lentils
- ¼ cup of rice
- 1 potato
- 1 red pepper
- 1 onion
- 1 clove of garlic
- 4 Tbsp. of extra-virgin olive oil
- 2 tsp. of vinegar
- 1 bay leaf
- Paprika
- A pinch of salt

Directions

Put the lentils to cook in a pot of water with 1 Tbsp. of olive oil, salt and a bay leaf. Let it all boil for 30 minutes. Add 1 Tbsp. of vinegar, the rice, and the peeled and cubed potatoes into the pot. While the stew cooks, prepare the "sofrito" with the diced onion, pepper, and minced clove of garlic.

When the onion is transparent in color, turn off the heat and add a tsp. of paprika, remove from the heat, and pour the "sofrito" into the pot of lentils and rice.

Cook until the rice and lentils are fully cooked, about 90 minutes.

12.3.4 Greek "Fasolada"

Ingredients *Yields 4 servings*

- 1½ cups of dry white beans
- 2 red onions
- 2 big carrots
- 2 ripe tomatoes
- 1 stalk of celery
- 1 clove of garlic
- 1 bay leaf
- 1 cup of extra-virgin olive oil
- Salt

Directions

The day before, soak the white beans in cold water. The following day, in a pot of water covering all the beans, boil them for 40 minutes. Meanwhile, slice the carrots, chop the onions, and garlic. Peel the tomato and take out the seeds. Chop the celery.

Once the beans are cooked, drain them and add them back to the pot. Add the vegetables, the bay leaf, and salt. Cover the ingredients with water and add the olive oil. Let it cook at low heat for at least an hour. During this time, keep adding small amounts of water. Serve hot.

12.3.5   Sautéed Peas

Ingredients *Yields 4 servings*

- 3 cups of fresh or frozen peas
- 2 firm onions
- 1 clove of garlic
- 3.5 oz. of prosciutto
- 4 cups of water
- 2 tsp. of paprika
- Extra-virgin olive oil
- Salt

Directions

Chop the garlic and onions. Sauté them in a pot with olive oil. When they start to change color, add the ham cut into strips to the pot and continue to cook the "sofrito" for another minute. Then add the paprika to the pot, stir a few times, and quickly add a little water so that it does not burn. Add the peas and cover them with the rest of the water. Let it cook until the peas are soft.

12.4   EGG

12.4.1   Spanish Omelet With "Pisto"

Ingredients *Yields 4 servings*

Spanish omelet

- 4 eggs
- 4 potatoes
- Extra-virgin olive oil
- Salt

Pisto

- 1 red tomato
- 1 green pepper
- 1 red pepper
- 1 onion
- 2 cloves of garlic

- Extra-virgin olive oil
- Salt

Directions

Peel the potatoes and cut them into thin slices. Fry them in olive oil until they turn golden and then drain them. Beat the eggs, salt to taste, add the slices of potato, and mix well. Cook the omelet on both sides at low heat until it has completely cooked.

To make the side dish "pisto," chop the onion, peppers, tomato, and garlic into small squares. Put the cut vegetables to poach in a small pot with a little olive oil at very low heat until they soften. Salt to taste and let them cook at medium heat for 20 minutes stirring frequently. Serve the Spanish omelet with the side vegetables.

12.4.2   Broccoli and Cheese Omelet

Ingredients *Yields 4 servings*

- 1¼ cups of cooked broccoli
- ½ tomato
- ½ onion
- 4 eggs
- ¼ cup of shredded mozzarella
- 3 Tbsp. of milk
- Black ground pepper
- Extra-virgin olive oil
- Salt

Directions

Beat the eggs with the cheese, milk, salt, and pepper. Cut the onion and tomato into small squares. In a pan with olive oil, sauté the onion and tomato at low heat for 10 minutes. Meanwhile, preheat the oven to 300°F. When the "sofrito" is ready add it to the eggs and lastly, add the cooked broccoli. Grease a cooking dish with a little oil and pour the mixture. Bake for 20 minutes and remove from the baking dish. It can be served alone or with tomato sauce as a side dish.

12.4.3   Scrambled Eggs With Asparagus

Ingredients *Yields 4 servings*

- 6 eggs
- 1½ cups of white asparagus, chopped

- ½ onion
- 2 cloves of garlic
- 3 Tbsp. of extra-virgin olive oil
- A pinch of salt

Directions

Wash and chop the white asparagus into small pieces of equal length. In a saucepan with boiling water and salt, cook them for a minute to lose their firmness. Afterward, drain them and set them aside. Cut the onion into thin long strips and the garlic into thin slices and sauté them with olive oil. When the onion is transparent in color, add the asparagus and cook them together for a few minutes. Add the eggs to the pan and stir them carefully until they are partially cooked, not completely cooked. Serve immediately so that they do not become dry.

12.4.4 Poached Eggs With Clams and Marinara Sauce

Ingredients *Yields 4 servings*

- 4 eggs
- 3.5 oz. of clams
- ½ cup of white wine
- 1 onion
- ¼ cup of leeks, chopped
- 2 cloves of garlic
- ¼ cup of tomato sauce
- 3 Tbsp. of extra-virgin olive oil
- 2 tsp. of flour
- Parsley
- 1 Tbsp. of vinegar
- A pinch of salt

Directions

Place the cleaned clams into a pot with a little bit of white wine on the stove. When they open, drain the liquid and set it aside for the marinara sauce.

Preparing the Marinara Sauce:

Dice the onion, garlic cloves, and leek. Poach the vegetables and once poached, add a little flour and tomato sauce.

Sauté all the vegetables with the sauce together and add the liquid from the clams from when they were cooked. Sprinkle with a little parsley and add salt to the sauce.

Take out the clams from their shells and mix with the marinara sauce.

To poach the eggs, use a wide pot with some depth and boil water. Add a pinch of salt and a drizzle of vinegar. Crack the eggs and place them one by one into the boiling pot of water. With the help of a skimmer, wrap the egg white around the yolk. Take the eggs out when the egg whites are cooked around the yolks. Place them over the marinara sauce and serve hot.

12.4.5 Fried Eggs With Mushrooms

Ingredients *Yields 4 servings*

- 4 eggs
- Extra-virgin olive oil for frying
- 1½ cups of mushrooms (10.5 oz.)
- 1 leek
- 4 Tbsp. of extra-virgin olive oil
- A pinch of salt

Directions

Wash the mushrooms and cut them into four pieces. Cut the leek into thin long strips and lightly sauté them in a pot with olive oil. When the leeks are sautéed, add the mushrooms, add salt to taste, and cook at medium heat together with the leeks for about 20 minutes. In a pan with heated olive oil, crack the eggs one by one and fry them in the traditional way, that is, with very hot oil so that the borders of the egg whites golden while the yolk remains liquid. Add a pinch of salt before taking the eggs out of the pan with a skimmer and serve with the sautéed mushrooms.

12.5 FISH

12.5.1 Sole With Clams

Ingredients *Yields 4 servings*

- 8 fillets of sole
- ½ lb. of clams
- ¼ cup of flour
- ¾ cup of onion, diced
- 2 Tbsp. of extra-virgin olive oil
- 1 clove of garlic
- Parsley
- A pinch of salt

Directions

Place the clams into a bowl filled with cold water and salt to clean the sand off. Dice the onion and garlic. Heat the olive oil in a low pot and poach the onion. As the onion starts to become transparent in color, add the garlic and place the previously salted fillets of sole into the pot. Add the flour, turn over the fillets a few times, and add cold water a little at a time until they are covered. The sauce should just cover the fish. In a separate pot, open the clams by steaming them. Throw away the ones that do not open. Add the open clams to the pot and stir for a few minutes to mix with the sauce. Finish by garnishing with chopped parsley.

12.5.2   Squid Skewers

Ingredients *Yields 4 servings*

- ½ lb. of squid
- ½ lb. of salmon
- 1 green pepper
- 1 Tbsp. of extra-virgin olive oil
- A pinch of salt

Directions

Cut the squid and the salmon into thick squares and the green pepper into pieces. Make the skewers by alternating the fish with pieces of green pepper. Brush a layer of olive oil over the skewers and add a pinch of salt.

On a nonstick pan or a grill let the skewers golden on both sides and then lower the heat so that the fish cooks inside.

Mayonnaise can be served with the skewers if desired.

12.5.3   Codfish Served With Onion and Tomato Sauce

Ingredients *Yields 4 servings*

- 1.8 lbs. of fresh codfish
- 2 cups of onion, chopped
- ¼ cup of tomato sauce
- Extra-virgin olive oil
- Flour
- Parsley
- A pinch of salt

Directions

Cut the codfish into pieces and cover with flour. Fry in a pan at medium heat with olive oil and set aside.

Cut the onion in thin strips and cook at low heat in a pan with a little olive oil and salt. Once the onion is poached, add the tomato sauce and salt to taste.

Place the codfish over the bed of onion and cook a few minutes all together. Finish by topping off with minced parsley.

12.5.4   Grilled Salmon With Rice

Ingredients *Yields 4 servings*

- 4 salmon fillets
- 4 cloves of garlic
- 3 Tbsp. extra-virgin olive oil
- A pinch of salt

  For the side dish

- ½ cup of rice
- ½ onion
- 1 green pepper
- 1 red pepper
- 1 clove of garlic
- 3 Tbsp. of extra-virgin olive oil
- A pinch of salt

Directions

Peel the cloves of garlic and mash it (crushing it in a mortar or cutting it very finely with a mincer) together with the parsley.

To this mixture add a little olive oil, a pinch of salt, and spread onto the salmon fillets.

Cook the fish on a pan or nonstick grill for 3 minutes on each side with a little olive oil as a base.

Set aside on a platter. To prepare the side dish, cut the onion and peppers in squares. Heat oil in a pot and sauté the onion at low heat until it begins to golden.

Add the mashed garlic and let it cook for 5 more minutes. Add a pinch of salt. Boil water in a pot, add the rice, and cook for about 20 minutes. Once it has cooked, drain and add the sautéed vegetables. Sauté the combination for a couple minutes and serve with the salmon fillets.

12.5.5 Swordfish With Tomato and Black Olives

Ingredients *Yields 4 servings*

- 4 swordfish fillets
- 1 lb. of tomatoes
- 4 cloves of garlic
- ½ onion
- 12 pitted black olives
- Extra-virgin olive oil
- Salt

Directions

Dice the onion and cloves of garlic. Grate the tomatoes. In a pan with a little olive oil, sauté the onion and garlic, add the tomatoes and cook at low heat for a few minutes. Salt the swordfish and cook in a pan with oil until they are cooked on each side. Serve the fish with the tomato sauce on top and the olives cut into slices.

12.6 MEATS

12.6.1 Baked Chicken With Vegetables

Ingredients *Yields 4 servings*

- 1 chicken
- 3 potatoes
- 4 onions
- 1 zucchini
- Extra-virgin olive oil
- Lemon juice
- A pinch of salt

Directions

Preheat the oven to 355°F.

Peel the potatoes and cut in half. Peel the onions and cut into strips. Clean the zucchini and cut into square pieces. Place all the vegetables in an oven safe platter, add salt and a drizzle of olive oil.

Mix the olive oil with lemon juice and coat the chicken inside and out. Place the chicken on a new cooking platter and put both platters into the oven for about 1 hour until the chicken has baked. Take the platters out of the oven and cut the chicken. Serve with the baked vegetables.

12.6.2 Chicken Thighs With Olives

Ingredients *Yields 4 servings*

- 4 boneless chicken thighs
- ½ cup of green olives
- 1 onion
- 1 clove of garlic
- ½ cup of white wine
- 3 Tbsp. of extra-virgin olive oil
- 1 Tbsp. of flour
- Basil
- A pinch of salt

Directions

Flour the chicken thighs and lightly fry in a low pot with 3 Tbsp. of olive oil.

Remove from the heat and set aside to continue cooking with later.

Next, cut the onion and garlic very finely and in the same pot cook the chicken at medium heat with the remaining olive oil until golden.

Put the chicken thighs in the pot, add the wine, and cover the pot so that the meat cooks at low heat.

Let the wine evaporate almost completely.

Add the olives (half of them in halves and the other half cut into small pieces) to the pot and continue to cook with the top on at medium heat for another 30 minutes.

Add a drizzle of water or chicken broth if the thighs become too dry.

Serve the thighs with the olive sauce and olives on the side.

12.6.3 Turkey Cooked With Tomatoes and Peppers

Ingredients *Yields 4 servings*

- 1 lb. of turkey pieces
- 4 ripe tomatoes
- 2 peppers
- 1 onion
- 1 clove of garlic
- 4 oz. of prosciutto
- ½ cup of white wine
- Black pepper
- Extra-virgin olive oil
- Salt

Directions

Golden the turkey in a pot with olive oil. Take out and set aside. In the same pot sauté the garlic and diced onion. When the onion is poached, add the prosciutto cut into small cubes and add the peppers cut into strips. Sauté for a few minutes. Add the diced tomatoes and the white wine. Lastly, add the turkey to the pot and cook all together for 15—20 minutes.

12.6.4   Turkey With Sautéed Zucchini and Onion Sandwich

Ingredients *Yields 4 servings*

- 8 slices of whole-grain bread
- 1 lb. of turkey breast
- 1 zucchini
- 1 onion
- 1 tomato
- 3 Tbsp. of extra-virgin olive oil
- A pinch of salt

Directions

Peel and cut the onion and zucchini into thin slices and sauté them in a pan with a little olive oil.

When the vegetables begin to soften, add thinly filleted turkey breasts, to cook on either side for 1 minute each.

Set aside, away from the heat, with the pan covered and wait for it to cool.

To make the sandwiches, create a base of thinly sliced tomatoes on a slice of bread and place the turkey breasts on top. Spread the cooked zucchini and onion on the turkey. Top with another slice of bread and serve warm.

12.6.5   Pork Tenderloin With Apple Sauce

Ingredients *Yields 4 servings*

- 8 pieces of pork tenderloin
- 4 golden apples
- Extra-virgin olive oil
- Salt

Directions

Grease the sliced pork tenderloin on both sides with olive oil and cook on the pan until golden. Add salt to taste. To prepare the apple sauce, peel and cut the apples into four pieces, take out the core, and then cut them into smaller pieces. Put the pieces of apple in a pot at medium heat and stir them occasionally with a wooden spoon. Cook for 30 minutes. When the sauce is cooked, set it aside so it cools and afterward put it through a strainer. Serve two pieces of pork on each plate with the apple sauce on the side.

Index

Note: Page numbers followed by "*f*" and "*t*" refer to figures and tables, respectively.

A

Alcohol consumption, aspects of
  cardiovascular diseases and, 153
  depression and, 154
  health outcomes, 155
  incidence of diabetes, 154
  mortality, 152–153
  recommendations, 155–156
  risk of cancer, 154–155
  risk of obesity or weight gain, 154
Alpha-linolenic acid (ALA), 40, 40*f*, 41*f*, 47,
  133–134
Alternate Mediterranean Diet (aMED), 11
Alzheimer's disease, 170–171

B

Binge-drinking, 151–152
Biological redundancy, 171

C

Carbohydrate intake, 36
Cardiovascular diseases
  alcohol consumption and, 153
  effect of climatic conditions, 161–162
  fish and, 133, 139–140
  fruits and vegetables and, 102–105
  legumes and, 118–119
  olive oil and, 66–82, 67*t*
  physical activity and, 161–162
  red and processed meat and, 141–142
Cardiovascular risk factors in Mediterranean
  countries, 160
Carotenes, 59–60
Case-control studies, 31–32, 32*f*
Cereals, 3
  in Mediterranean Diet, 111–112
  whole-grain, 111. *See also* Whole-grain
    cereals

Cheese, 3, 52
Cholesterol, 39
  HDL-C, 39, 51
    effects of olive oil, 61–63
  LDL-C, 39, 44–45, 51, 122
    effects of olive oil, 61–63
Chylomicron, 42–44, 43*f*
Cohort studies, 31–33, 33*f*
Cretan diet, 101
Cross-sectional studies, 31–32
Cumulative incidence, 27

D

DASH (Dietary Approaches to Stop
    Hypertension), 171–172
Dementia, 169–170
  common causes of, 170
    inflammation, 170–171
    oxidative stress, 170
  Mediterranean Diet and, 171–172,
    180–181
    epidemiological evidences, 172–180,
      173*t*
    LIILAC study, 180–181
    PATH Through Life Study, 178–179
    PREDIMED-PLUS trial, 180–181
    PREDIMED Study, 179–180
    recommendations, 181–182
Depression
  alcohol consumption and, 154
  cardiovascular disease and,
    184–185
  etiological hypothesis of, 182–184
  Mediterranean Diet and, 185
    epidemiological evidences, 185–191,
      186*t*, 187*t*
    recommendations, 191–192
Desaturase, 40
Dietary fats, 35, 37
  consumption of daily, 35

Dietary fibers, 121–122
Dietary Guidelines for Americans 2015, 13–14, 53–54
Dietary saturated fats, 51–52
Docosahexaenoic acid (DHA), 40*f*, 133–134

E
Eating habits in Mediterranean countries, 2
Ecologic studies, 31
Eggs, 3, 7
Eicosapentaenoic acid (EPA), 40*f*, 133–134
Empty calories, 154
Epidemiological studies, 30–34
   characteristics, 30–31
Essential fatty acids, 40, 40*f*
EUROLIVE dietary intervention study, 63
Experimental studies, 30
Exposure
   assignment, 30
   measures of association, 28–30
   nutritional epidemiology, classifying to, 25–26
Extra virgin olive oil, 45
   chemical composition of, 46*t*
   effect on cardioprotective function, 62*t*
   phenolic compounds of, 47*t*

F
Fat hardening, 49
Fat(s)
   absorption, 42–44, 43*f*
   bad, 48–51
   consumption, 52–55
   definition, 37–42
   deposited in adipocytes, 35
   dietary, 35, 37
   functions of, 35
   good, 37, 44–48
   healthy and unhealthy, 54*f*
   in-between, 51–52
   recommendation to reduce intake, 36
   solubility, 35
   -soluble vitamins, 35
Fatty acids, 38*f*, 39*f*
Fatty fish, 47
Fenugreek *(Trigonella foenum graecum)*, 122, 126
Fiber, 89
First World Conference on the Mediterranean Diet, 8

Fish, 3, 7
   adverse effects of mercury from, 142–143
   consumption, benefits of, 48
   intake and reduced risk of cardiovascular disease, 133, 139–140
   intermediate physiological effects of, 135–137
   nutritional composition, 133–134
   recommendations for intake, 143–144
Fish contaminants, 48
Flavonoids, 125
Follow-up time, 31
Fruits and vegetables, 101
   average serving of fruit, 106
   average serving of vegetables, 106
   against cardiovascular risk factors, 103–105
      hypertension, 104
      obesity, 103–104
      type 2 diabetes, 105
   protection against cardiovascular disease, 102–103
   recommended intake, 105–106
   servings for every day, 107
   tips to consume, 107

G
Grains, 3

H
Harvard T.H. Chan School of Public Health, 4–5
Hazard function, 28
Hazard ratios, 30
Healthy Diet Index, 171–172
Herbs, 4
High-fat diets, 54–55
Hydrogenated fatty acids, 41
Hydrogenation, 48–50
Hyperglycemia, 138
Hyperinsulinemia, 138
Hypertension
   management
      by fruits and vegetables, 104
      by olive oil, 79–80
   meat intake and, 139

I
Incidence rate or incidence density, 27–28
Individual studies, 31

Insulin resistance, 138
International Conference on the Diets of the
        Mediterranean organized by Oldways
        Preservation & Exchange Trust, 4−5
International Mediterranean Diet Foundation, 8

L

Legumes, 3, 6
    antioxidants, 126
    beneficial effects, 117
    constituents of, 117−118
    consumption and production of, worldwide,
        118
    fibers, 121−122
    flavonoids, 125
    intake and cardiovascular disease,
        epidemiological evidence, 118−119
        Costa Rican study, 118−119
        Japan Collaborative Cohort Study, 118
        National Health and Nutrition
            Examination Survey Epidemiologic
            Follow-up Study, 118
        PREDIMED trial, 118
    low glycemic index of, 122
    in Mediterranean Diet, 117−118
    nutrient content of, 119−126
        macronutrients, 120t
        of selected beans, 121t
    physiological effects of, 119−120
    phytochemical compounds, 119−126, 123t
    phytosterols, 125
    polyphenolic compounds, 124−125
    proteins, 120
    reduction of CVD risk factors, 119
    saponins, 125−126
    starch content of beans, 121−122
Linoleic acid, 35−36, 40, 40f, 41f, 47
Linolenic acid, 35−36
Lipids, 35
    compounds, 38
    derivatives, 38
    simple, 38
Lipoproteins, 39
Longitudinal studies, 31
Low-fat diets, 54−55

M

Measures of association, 28−30, 29f
    hazard ratios, 30
    odds ratio, 29−30
    relative risk, 28−29

Meat, 3
Mediterranean Adequacy Index (MAI), 12−13
Mediterranean alcohol-drinking, 152
Mediterranean countries
    cardiovascular risk factors in, 160
    climate conditions and opportunities for a
        healthy lifestyle, 161
    climatic conditions, 161−162
    degree of socialization in, 163−164
    Mediterranean lifestyle, 163
    sleep pattern characteristic of, 163−164
    vitamin D intake in, 161
Mediterranean Diet, 44, 54−55, 90, 101,
        106−107, 151−152, 159−160
    adherence to, 11−13, 187t
    as an intangible and sustainable food
        culture, 2
    beneficial health effects of, 1, 185−191
        CARDIVEG study, 17
        Chicago Health and Aging Project, 190
        Doetinchem Cohort Study, 15
        EPIC-Greece study, 15
        EPIC-Netherland study, 15
        EPIC-Spain study, 15
        epidemiological evidences, 14−21
        Health Alcohol and Psychosocial factors
            in Eastern Europe (HAPIEE) study, 15
        Lyon Diet Heart Study, 16−17
        MedLey study, 191
        MooDFOOD prevention trial, 191
        Northern Manhattan Study, 16
        Nurses' Health Study (NHS), 16
        PREDIMED trial, 17, 18t
        SMILES study, 191
        SUN project, 186
        Swedish Mammography Cohort, 15
        Women's Health Initiative study, 16
    cardio-protective effect of, 185
    cereals in, 111−112
        PREDIMED trial, 111−112
    characteristics, 2−4
    definitions and history of, 1−2
    dementia and, 169−182
    depression and, 182−192
    legumes in, 117−118
    main and typical foods of the traditional
        diet of, 3−4
    nutritional composition and nutritional
        adequacy of, 8−10
    pyramids, 4−8, 4f, 6f, 7f, 9f
        dietary pattern between 2009 and 2010, 5
        levels of recommended consumption, 5

Mediterranean Diet (*Continued*)
  physical activity, 5
  recommendations for consumption, 5—6
  weekly consumption pattern, 6—7
  wine consumption, 5
Mediterranean Diet 55 Score (MD55), 12
Mediterranean Diet Score (MDS), 11, 186
Mercury from fish, adverse effects of,
  142—143
Meta-analysis, 34
Micelles, 42
Mixed dementia, 170
MONICA Project (MONItoring
  CArdiovascular disease), 66
Monounsaturated fatty acids, 40, 44—46,
  59—60, 62—63, 65—66, 74, 90—92, 137

N

Neutrophil gelatinase-associated lipocalin
  (NGAL), 184
Nonsoy legumes, 119
Normann, Wilhelm, 49
Nutritional epidemiology, classifying exposure
  to, 25—26, 26*f*
Nuts, 3
  effects on health, 90—95
    clinical trials, 94—95
    fat malabsorption, 95
    observational studies, 92—94
    PREDIMED trial, 95
    satiety power, 95
    waist circumference and body mass
      index, 95
  incorporation in diet, 96—97
  mechanisms of cardiometabolic protection
    exerted by, 92*t*
  nutrient content of, 89—92, 91*t*
  recommendations for intake, 96—97
  source of L-arginine, 89

O

Obesity management
  by fruits and vegetables, 103—104
  by olive oil, 81
Observational studies, 30
Occurrence of a disease, quantifying, 27*f*
  cumulative incidence, 27
  hazard function, 28
  incidence rate or incidence density, 27—28
  prevalence odds, 26
  prevalence proportion, 26

Odds ratio, 29—30
Olea europea L., 60
Oleic acid, 44—45, 62—63
Oleuropein, 59—60
Olive oil
  cardiovascular risk factors and,
    epidemiological studies, 74—82, 75*t*
    diabetes, 74—79
    EPIC-Interact study, 78
    hypertension, 79—80
    metabolic syndrome, 81—82
    obesity, 81
    PREDIMED trial, 74, 78—82
    SUN cohort study, 74, 78, 81
  composition of, 59—60, 60*t*
  consumer tips, 82—83
  effects on health, 61—66
    antioxidant and antiinflammatory effect,
      63—64
    antithrombotic effect, 65—66
    cardiovascular health, 66—73, 67*t*
    endothelial function, effects on, 64—65
    LDL and HDL cholesterol, effects on,
      61—63
  incorporating in diet, 83
  pomace, 61
  pure, 61
  recommended rate of intake, 82
  virgin, 60—61
Olives and olive oil, 3
Omega-3 consumption, 47—48
Omega-3 fatty acids, 40*f*
Omega-3 PUFAs, 134—135
  health effects
    antiarrhythmia effects, 135, 136*f*, 137
    effects on triglycerides, 135
    heart rate and blood pressure, 136
    inflammatory responses, 137
    myocardial filling and efficiency,
      136—137
    visceral adiposity, 137
Oxidative stress, 45

P

Pancreatic lipase, 42
Parkinson's disease, 171
Partially hydrogenated oil, 48—49
Phenolic compounds, 59—60
Phospholipids, 38—39
Phytic acid, 124
Phytochemical compounds, 122—126, 123*t*
Phytochemicals, 106—107

Phytosterols, 125
14-Point Mediterranean Diet Adherence
    Screener (MEDAS), 12
Polyphenolic compounds, 124−125
Polyphenols, 46, 65−66
Polyphenols, 60−61
Polyunsaturated fatty acids, 40−41, 47−48,
    59−60, 62, 74
Potatoes, 6
PREDIMED trial, 63, 65
Prevalence odds, 26
Prevalence proportion, 26
Processed meats, 7
Prospective studies, 31
Pulses, 117

R
Randomized controlled trials, 33, 33*f*, 90
Recipes
    baked chicken with vegetables
        directions, 213−214
        ingredients, 213
    broccoli and cheese omelet
        directions, 208−209
        ingredients, 208
    chicken thighs with olives
        directions, 214
        ingredients, 214
    chickpea and egg salad
        directions, 205
        ingredients, 204−205
    chickpeas with codfish
        directions, 205−206
        ingredients, 205
    codfish served with onion and tomato sauce
        directions, 212
        ingredients, 211−212
    cream of butternut squash, 201
        directions, 201
        ingredients, 201
    "Escalibada" toast
        directions, 202
        ingredients, 201−202
    feta cheese and nut spinach salad, 200
        directions, 200−201
        ingredients, 200
    fried eggs with mushrooms
        directions, 210
        ingredients, 210
    Greek "Fasolada"
        directions, 206−207
        ingredients, 206

grilled salmon with rice
    directions, 212−213
    ingredients, 212
Mediterranean lentils
    directions, 206
    ingredients, 206
orange and raisin endive salad, 200
    directions, 200
    ingredients, 200
pasta with red peppers and tuna
    directions, 202−203
    ingredients, 202
pea and mushroom spaghetti
    directions, 203
    ingredients, 202−203
poached eggs with clams and marinara
    sauce
    directions, 209−210
    ingredients, 209
pork tenderloin with apple sauce
    directions, 215−216
    ingredients, 215
sautéed peas
    directions, 207
    ingredients, 207
scrambled eggs with asparagus
    directions, 209
    ingredients, 208−209
seafood paella
    directions, 204
    ingredients, 204
sole with clams
    directions, 211
    ingredients, 210−211
Spanish omelet with "pisto"
    directions, 208
    ingredients, 207
squid skewers
    directions, 211
    ingredients, 211
swordfish with tomato and black olives
    directions, 213
    ingredients, 213
tomato soup, 201
    directions, 201−202
    ingredients, 201
turkey cooked with tomatoes and peppers
    directions, 215
    ingredients, 214−215
turkey with sautéed zucchini and onion
    sandwich
    directions, 215

Recipes (*Continued*)
    ingredients, 215
    vegetable soup with brown rice
        directions, 204
        ingredients, 203–204
    whole grain noodles with mussels
        directions, 203
        ingredients, 203
Red and processed meat, 7
    intake and reduced risk of cardiovascular
        disease, 141–142
    intermediate physiological effects of,
        137–139
        arterial compliance and vascular
            stiffness, 139
        body weight, 138–139
        hypertension, 139
        irregularities in glucose metabolism,
            138
        LDL:HDL serum cholesterol ratio,
            138
        triglyceride levels, 138
    nutritional composition of, 134–135
Red wine, 151. *See also* Alcohol
    consumption, aspects of
Refined carbohydrates, 36, 53–54
Relative risk, 28–29
Retrospective studies, 31

S

Sabatier, Paul, 49
Saponins, 125–126
Saturated fatty acids, 39, 51–52, 59–60, 62,
    74
Seeds, 3
Semi essential fatty acids, 40
Seven Countries Study, 1
Shellfish, 3
Socialization, 7
Spices, 4
Squalene, 59–60
Sweet foods, 7

T

Tissue factor pathway inhibitor (TFPI),
    65–66
Tocopherols, 59–60
Trans fat, 48–51
    adverse effects, 50
    heart disease and, 49–50
Triglycerides, 37, 37*f*, 41–42
    effects of olive oil, 61–63

U

UNESCO's Representative List of Intangible
    Cultural Heritage of Humanity, 2
Unsaturated fatty acids, 39
Urban environments and their relationship to
    food consumption, 165–166, 165*f*

V

Vascular dementia, 170
Virgin olive oil, 60–61
Vitamins A, D, E, and K, 35, 60–61, 89
Volatile compounds, 59–60

W

White meat, 7
Whole-grain cereals, 111
    biological mechanisms of benefit of,
        115–116
    cardiovascular disease, epidemiological
        evidence, 112–115
    recommendations for the consumption,
        116–117
WHO MONICA Project, 159
Wine, 4
World Health Organization (WHO)/Food and
    Agriculture Organization (FAO), 4–5

Y

Yogurt, 3